POPULATION

POPULATION

An Introduction to Concepts and Issues
Twelfth Edition

John R. Weeks
San Diego State University

CENGAGE Learning®

Australia • Brazil • Japan • Korea • Mexico • Singapore • Spain • United Kingdom • United States

Population: An Introduction to Concepts and Issues, Twelfth Edition

John R. Weeks

Product Director: Marta Lee-Perriard

Product Manager: Jennifer Harrison

Content Developer: Liana Sarkisian

Product Assistant: Julia Catalano

Media Developer: John Chell

Marketing Manager: Kara Kindstrom

Production Management, and Composition: Manoj Kumar, MPS Limited

Art Director: Caryl Gorska

Manufacturing Planner: Judy Inouye

Text Researcher: Nandhini Srinivasagopalan, Lumina Datamatics

Manuscript Editor: Deanna Weeks

Copy Editor: MPS Limited

Cover Designer: Caryl Gorska

Library of Congress Control Number: 2014947006

ISBN: 978-1-305-09450-5

Cengage Learning
20 Channel Center Street
Boston, MA 02210
USA

Cengage Learning is a leading provider of customized learning solutions with office locations around the globe, including Singapore, the United Kingdom, Australia, Mexico, Brazil, and Japan. Locate your local office at **www.cengage.com/global.**

Cengage Learning products are represented in Canada by Nelson Education, Ltd.

To learn more about Cengage Learning Solutions, visit **www.cengage.com.**

Purchase any of our products at your local college store or at our preferred online store **www.cengagebrain.com.**

Printed in the United States of America
Print Number: 01 Print Year: 2014

To Deanna

BRIEF TABLE OF CONTENTS

PART ONE **A DEMOGRAPHIC PERSPECTIVE**
CHAPTER 1 Introduction to Demography 1
CHAPTER 2 Global Population Trends 25
CHAPTER 3 Demographic Perspectives 58
CHAPTER 4 Demographic Data 100

PART TWO **POPULATION PROCESSES**
CHAPTER 5 The Health and Mortality Transition 139
CHAPTER 6 The Fertility Transition 189
CHAPTER 7 The Migration Transition 251

PART THREE **POPULATION STRUCTURE AND CHARACTERISTICS**
CHAPTER 8 The Age Transition 298
CHAPTER 9 The Urban Transition 343
CHAPTER 10 The Family and Household Transition 384

PART FOUR **USING THE DEMOGRAPHIC PERSPECTIVE**
CHAPTER 11 Population and Sustainability 432
CHAPTER 12 What Lies Ahead? 481

GLOSSARY 505
BIBLIOGRAPHY 519
GEOGRAPHIC INDEX 549
SUBJECT INDEX 559

DETAILED TABLE OF CONTENTS

PREFACE xxi

PART ONE
A DEMOGRAPHIC PERSPECTIVE

CHAPTER 1
INTRODUCTION TO DEMOGRAPHY 1

WHAT IS DEMOGRAPHY? 3

HOW DOES DEMOGRAPHY CONNECT THE DOTS? 5
The Relationship of Population to Resources 6
The Relationship of Population to Social and Political Dynamics 7
ESSAY: *Demographic Contributions to the "Mess in the Middle East"* 10
How Is the Book Organized? 21

SUMMARY AND CONCLUSION 22

MAIN POINTS 22

QUESTIONS FOR REVIEW 23

WEBSITES OF INTEREST 23

CHAPTER 2
GLOBAL POPULATION TRENDS 25

WORLD POPULATION GROWTH 26
A Brief History 26
How Fast Is the World's Population Growing Now? 29
The Power of Doubling—How Fast Can Populations Grow? 30
Why Was Early Growth So Slow? 31
Why Are More Recent Increases So Rapid? 32

How Many People Have Ever Lived? 34
Redistribution of the World's Population through Migration 34

GEOGRAPHIC DISTRIBUTION OF THE WORLD'S POPULATION 37

GLOBAL VARIATION IN POPULATION SIZE AND GROWTH 39
North America 40
Mexico and Central America 42
South America 43
Europe 44
ESSAY: *Implosion Or Invasion? The Choices Ahead For Low-Fertility Countries* 46
Northern Africa and Western Asia 48
Sub-Saharan Africa 49
South and Southeast Asia 50
East Asia 52
Oceania 54

SUMMARY AND CONCLUSION 55

MAIN POINTS 55

QUESTIONS FOR REVIEW 56

WEBSITES OF INTEREST 57

CHAPTER 3
DEMOGRAPHIC PERSPECTIVES 58

PREMODERN POPULATION DOCTRINES 61

THE PRELUDE TO MALTHUS 66

THE MALTHUSIAN PERSPECTIVE 67
Causes of Population Growth 68
Consequences of Population Growth 70
Avoiding the Consequences 70
Critique of Malthus 71
Neo-Malthusians 73
ESSAY: *Who are the Neo-Malthusians?* 74

THE MARXIAN PERSPECTIVE 76
Causes of Population Growth 76
Consequences of Population Growth 76
Critique of Marx 77

THE PRELUDE TO THE DEMOGRAPHIC TRANSITION THEORY 79
Mill 79
Dumont 80
Durkheim 81

THE THEORY OF THE DEMOGRAPHIC TRANSITION 81
Critique of the Demographic Transition Theory 84
Reformulation of the Demographic Transition Theory 85
The Theory of Demographic Change and Response 88
Cohort Size Effects 90
Is There Something Beyond the Demographic Transition? 91

THE DEMOGRAPHIC TRANSITION IS REALLY A SET OF TRANSITIONS 92
The Health and Mortality Transition 93
The Fertility Transition 93
The Age Transition 94
The Migration Transition 94
The Urban Transition 95
The Family and Household Transition 95
Impact on Society 96

SUMMARY AND CONCLUSION 97

MAIN POINTS 98

QUESTIONS FOR REVIEW 98

WEBSITES OF INTEREST 99

CHAPTER 4
DEMOGRAPHIC DATA 100

SOURCES OF DEMOGRAPHIC DATA 101
Population Censuses 101
The Census of the United States 105
Who Is Included in the Census? 110
Coverage Error 111
ESSAY: *Demographics of Politics: Why the Census Matters* 112
Measuring Coverage Error 116
Content Error 117
Sampling Error 118
Continuous Measurement—American Community Survey 118
The Census of Canada 119
The Census of Mexico 120
IPUMS—Warehouse of Global Census Data 121

REGISTRATION OF VITAL EVENTS 122

COMBINING THE CENSUS AND VITAL STATISTICS 125

ADMINISTRATIVE DATA 126

SAMPLE SURVEYS 127
Demographic Surveys in the United States 127
Canadian Surveys 128

Mexican Surveys 128
Demographic and Health Surveys 128
Demographic Surveillance Systems 129
European Surveys 129

HISTORICAL SOURCES 130

SPATIAL DEMOGRAPHY 131
Mapping Demographic Data 132
GIS and the Census 134

SUMMARY AND CONCLUSION 135

MAIN POINTS 136

QUESTIONS FOR REVIEW 137

WEBSITES OF INTEREST 137

PART TWO
POPULATION PROCESSES

CHAPTER 5
THE HEALTH AND MORTALITY TRANSITION 139

DEFINING THE HEALTH AND MORTALITY TRANSITION 140

HEALTH AND MORTALITY CHANGES OVER TIME 141
The Roman Era to the Industrial Revolution 142
The Industrial Revolution to the Twentieth Century 143
World War II as a Modern Turning Point 146
Postponing Death by Preventing and Curing Disease 148
The Nutrition Transition and Its Link to Obesity 149

LIFE SPAN AND LONGEVITY 150
Life Span 151
Longevity 151

DISEASE AND DEATH OVER THE LIFE CYCLE 153
Age Differentials in Mortality 153
Infant Mortality 154
Mortality at Older Ages 156
Sex and Gender Differentials in Mortality 158

CAUSES OF POOR HEALTH AND DEATH 159
Communicable Diseases 160
ESSAY: *Mortality Control and the Environment* 164
Noncommunicable Conditions 167
Injuries 168
The "Real" Causes of Death 168

MEASURING MORTALITY 170
Crude Death Rate 171
Age/Sex-Specific Death Rates 171
Age-Adjusted Death Rates 172
Life Tables 173
Life Table Calculations 173
Disability-Adjusted Life Years 180

HEALTH AND MORTALITY INEQUALITIES 180
Urban and Rural Differentials 181
Neighborhood Inequalities 181
Educational and Socioeconomic Differentials in Mortality 182
Inequalities by Race and Ethnicity 183
Marital Status and Mortality 184

SUMMARY AND CONCLUSION 185

MAIN POINTS 186

QUESTIONS FOR REVIEW 187

WEBSITES OF INTEREST 187

CHAPTER 6
THE FERTILITY TRANSITION 189

WHAT IS THE FERTILITY TRANSITION? 190

HOW HIGH COULD FERTILITY LEVELS BE? 191
The Biological Component 191
The Social Component 194

WHY WAS FERTILITY HIGH FOR MOST OF HUMAN HISTORY? 196
Need to Replenish Society 196
Children as Security and Labor 198
Lower Status of Women in Traditional Societies 198

THE PRECONDITIONS FOR A DECLINE IN FERTILITY 200

IDEATIONAL CHANGES THAT MUST TAKE PLACE 201
ESSAY: *Reproductive Rights, Reproductive Health, and the Fertility Transition* 202

MOTIVATIONS FOR LOWER FERTILITY LEVELS 205
The Supply-Demand Framework 205
The Innovation/Diffusion and "Cultural" Perspective 209

HOW CAN FERTILITY BE CONTROLLED? 211
Proximate Determinants of Fertility 213
Proportion Married—Limiting Exposure to Intercourse 213

Use of Contraceptives 215
Induced Abortion 219
Involuntary Infecundity from Breastfeeding 220
The Relative Importance of the Proximate Determinants 221

HOW DO WE MEASURE CHANGES IN FERTILITY? 223
Period Measures of Fertility 223
Cohort Measures of Fertility 229
Fertility Intentions 229

HOW IS THE FERTILITY TRANSITION ACCOMPLISHED? 230

GEOGRAPHIC VARIABILITY IN THE FERTILITY TRANSITION 232

CASE STUDIES IN THE FERTILITY TRANSITION 234
United Kingdom and Other European Nations 234
China 238
The United States 241

SUMMARY AND CONCLUSION 247

MAIN POINTS 248

QUESTIONS FOR REVIEW 249

WEBSITES OF INTEREST 250

CHAPTER 7
THE MIGRATION TRANSITION 251

WHAT IS THE MIGRATION TRANSITION? 252

DEFINING MIGRATION 253

INTERNAL AND INTERNATIONAL MIGRANTS 254

MEASURING MIGRATION 255
Stocks versus Flows 256
Migration Indices 258

THE MIGRATION TRANSITION WITHIN COUNTRIES 261
Why Do People Migrate? 262
Who Migrates? 266
Migration within the United States 267

MIGRATION BETWEEN COUNTRIES 268
Why Do People Migrate Internationally? 270
Who Migrates Internationally? 272

MIGRATION ORIGINS AND DESTINATIONS 274
Global Patterns of Migration 274
Migration into the United States 277

Migration out of the United States 283
ESSAY: *Is Migration A Crime? Illegal Immigration in Global Context* 284
Migration into and out of Canada 286

FORCED MIGRATION 288
Refugees and Internally Displaced Persons 288
Slavery 288

CONSEQUENCES OF MIGRATION 290
Consequences for Migrants 290
Children of Immigrants 292
Societal Consequences 292
Remittances 293

SUMMARY AND CONCLUSION 294

MAIN POINTS 296

QUESTIONS FOR REVIEW 296

WEBSITES OF INTEREST 297

PART THREE
POPULATION STRUCTURE AND CHARACTERISTICS

CHAPTER 8
THE AGE TRANSITION 298

WHAT IS THE AGE TRANSITION? 299

THE CONCEPTS OF AGE AND SEX 299
Age Stratification 300
Age Cohorts and Cohort Flow 301
Gender and Sex Ratios 304
The Feminization of Old Age 306

DEMOGRAPHIC DRIVERS OF THE AGE TRANSITION 307
The Impact of Declining Mortality 309
The Impact of Declining Fertility 313
Where Does Migration Fit In? 315

AGE TRANSITIONS AT WORK 317
The Progression from a Young to an Old Age Structure 317
Youth Bulge—Dead End or Dividend? 317
China's Demographic Dividend 318
What Happened to India's Demographic Dividend? 320
Demographic Dividends in the United States and Mexico 322

POPULATION AGING AS PART OF THE AGE TRANSITION 324
What Is Old? 324
ESSAY: *Who Will Pay for Baby Boomers to Retire in the Richer Countries?* 326

How Many Older People Are There? 328
Where are the Older Populations? 329
The Third Age (Young-Old) and Fourth Age (Old-Old) 332
Centenarians and Rectangularization—Is This the End of the Age Transition? 333

READING THE FUTURE FROM THE AGE STRUCTURE 334
Population Projections 334
Population Momentum 339

SUMMARY AND CONCLUSION 340

MAIN POINTS 341

QUESTIONS FOR REVIEW 342

WEBSITES OF INTEREST 342

CHAPTER 9
THE URBAN TRANSITION 343

WHAT IS THE URBAN TRANSITION? 344
Defining Urban Places 345

WHAT ARE THE DRIVERS OF THE URBAN TRANSITION? 347
Precursors 347
Current Patterns 349
The Urban Hierarchy and City Systems 352
An Illustration from Mexico 354
An Illustration from China 355

THE PROXIMATE DETERMINANTS OF THE URBAN TRANSITION 357
Internal Rural-to-Urban Migration 357
Natural Increase 358
International Urbanward Migration 362
Reclassification 362
Defining the Metropolis 363
ESSAY: *Nimby and Bnana—The Politics of Urban Sprawl in America* 364

THE URBAN EVOLUTION THAT ACCOMPANIES THE URBAN TRANSITION 368
Urban Crowding 369
Slums 371
Suburbanization and Exurbanization 374
Residential Segregation 377

CITIES AS SUSTAINABLE ENVIRONMENTS 379

SUMMARY AND CONCLUSION 381

MAIN POINTS 382

QUESTIONS FOR REVIEW 383

WEBSITES OF INTEREST 383

CHAPTER 10
THE FAMILY AND HOUSEHOLD TRANSITION 384

WHAT IS THE FAMILY AND HOUSEHOLD TRANSITION? 385
- Defining Family Demography and Life Chances 386
- The Growing Diversity in Household Composition and Family Structure 388
- Gender Equity and the Empowerment of Women 392

PROXIMATE DETERMINANTS OF FAMILY AND HOUSEHOLD CHANGES 393
- Delayed Marriage Accompanied by Leaving the Parental Nest 393
- Cohabitation 396
- Nonmarital Childbearing 397
- Childlessness 399
- Divorce 399
- Widowhood 400
- The Combination of These Determinants 401

CHANGING LIFE CHANCES 401
- Education 402
- Labor Force Participation 406
- Occupation 409
- Income 410
- Poverty 414
- Wealth 416
- **ESSAY:** *Show Me the Money! Household Diversity and Wealth Among the Elderly* 418
- Race and Ethnicity 420
- Religion 424

DOES MARRIAGE MATTER? 426

SUMMARY AND CONCLUSION 428

MAIN POINTS 429

QUESTIONS FOR REVIEW 430

WEBSITES OF INTEREST 431

PART FOUR
USING THE DEMOGRAPHIC PERSPECTIVE

CHAPTER 11
POPULATION AND SUSTAINABILITY 432

THE USE AND ABUSE OF THE EARTH'S RESOURCES 434
Economic Growth and Development 435
Measuring GNI and Purchasing Power Parity 436

HOW IS POPULATION RELATED TO ECONOMIC DEVELOPMENT? 439
Is Population Growth a Stimulus to Economic Development? 440
Is Population Growth Unrelated to Economic Development? 442
Is Population Growth Detrimental to Economic Development? 443

THE BOTTOM LINE FOR THE FUTURE: CAN BILLIONS MORE PEOPLE BE FED? 446
The Relationship between Economic Development and Food 446
Extensification—Increasing Farmland 449
Intensification—Increasing Per-Acre Yield 451
The Demand for Food Is Growing Faster Than the Population 456
How Many People Can Be Fed? 457

ENVIRONMENTAL DEGRADATION 461
Polluting the Ground 461
Polluting the Air 463
ESSAY: *How Big is Your Ecological Footprint?* 464
Damage to the Water Supply 468

HUMAN DIMENSIONS OF ENVIRONMENTAL CHANGE 469
Assessing the Damage Attributable to Population Growth 470
Environmental Disasters Lead to Death and Dispersion 471

SUSTAINABLE DEVELOPMENT—POSSIBILITY OR OXYMORON? 473
Are We Overshooting Our Carrying Capacity? 475

SUMMARY AND CONCLUSION 477

MAIN POINTS 478

QUESTIONS FOR REVIEW 479

WEBSITES OF INTEREST 480

CHAPTER 12
WHAT LIES AHEAD? 481

FROM REVOLUTION TO EVOLUTION 482
The Health and Mortality Evolution 484

The Fertility Evolution 487
The Migration Evolution 488
ESSAY: *A California Community Copes with the Migration Evolution* 490
The Age Evolution 493
The Urban Evolution 495
The Family and Household Evolution 496

WHAT CAN COUNTRIES DO TO INFLUENCE WHAT LIES AHEAD? 498

SUMMARY AND CONCLUSION 501

MAIN POINTS 502

QUESTIONS FOR REVIEW 503

WEBSITES OF INTEREST 503

GLOSSARY 505
BIBLIOGRAPHY 519
GEOGRAPHIC INDEX 549
SUBJECT INDEX 559

PREFACE

Growth, transition, and evolution. These are the key demographic trends as we move through the twenty-first century, and they will have huge impacts on your life. When I think about population growth in the world, I conjure up an image of a bus hurtling down the highway toward what appears to be a cliff. The bus is semiautomatic and has no driver in charge of its progress. Some of the passengers on the bus are ignorant of what seems to lie ahead and are more worried about whether the air conditioning is turned up high enough or wondering how many snacks they have left for the journey. Other more alert passengers are looking down the road, but some of them think that what seems like a cliff is really just an optical illusion and is nothing to worry about; some think it may just be a dip, not really a cliff. Those who think it is a cliff are trying to figure out how to apply the brakes, knowing that a big bus takes a long time to slow down even after the brakes are put on.

Are we headed toward a disastrous scenario? We don't really know for sure, but we simply can't afford the luxury of hoping for the best. The population bus is causing damage and creating vortexes of change as it charges down the highway, whether or not we are on the cliff route; and the better we understand its speed and direction, the better we will be at steering it and managing it successfully. No matter how many stories you have heard about the rate of population growth coming down or about the end of the population explosion (and those stories are true, up to a point), the world *will* continue to add billions more to the current 7 billion before it stops growing. Huge implications for the future lie in that growth in numbers.

The transitions represent the way in which population growth actually affects us. The world's population is growing because death rates have declined over the past several decades at a much faster pace than have birth rates, and as we go from the historical pattern of high birth and death rates to the increasingly common pattern of low birth and death rates, we pass through the demographic transition. This is actually a whole set of transitions relating to changes in health and mortality, fertility, migration, age structure, urbanization, and family and household structure. Each of these separate, but interrelated, changes has serious consequences for the way societies and economies work, and for that reason they have big implications for you personally. Over time, these transitions have evolved in ways that vary from one part of the world to another, and so their path and progress are less predictable than we once thought, but we have good analytical tools for keeping track of them, and potentially influencing them.

The growth in numbers (the bus hurtling toward what we hope is *not* a cliff) and the transitions and evolutions created in the process (the vortex created by the passing bus) have to be dealt with simultaneously, and our success as a human civilization depends on how well we do in this project. A lot is at stake here and my goal in this book is to provide you with as much insight as possible into the ways in which these demographic trends of growth, transition, and evolution affect your life in small and large ways.

Over the years, I have found that most people are either blissfully unaware of the enormous impact of population growth and change on their lives, or they are nearly overwhelmed whenever they think of population growth because they have heard so many horror stories about impending doom, or, increasingly, they have heard that population growth is ending and thus assume that the story has a happy ending. This latter belief is in many ways the scariest, because the lethargy that develops from thinking that the impact of population growth is a thing of the past is exactly what will lead us to doom. My purpose in this book is to shake you out of your lethargy (if you are one of those types), without necessarily scaring you in the process. I will introduce you to the basic concepts of population studies and help you develop your own demographic perspective, enabling you to understand some of the most important issues confronting the world. My intention is to sharpen your perception of population growth and change, to increase your awareness of what is happening and why, and to help prepare you to cope with (and help shape) a future that will be shared with billions more people than there are today.

I wrote this book with a wide audience in mind because I find that students in my classes come from a wide range of academic disciplines and bring with them an incredible variety of viewpoints and backgrounds. No matter who you are, demographic events are influencing your life, and the more you know about them, the better off you will be.

What Is New in This Twelfth Edition

Populations are constantly changing and evolving and each successive edition of this book has aimed to keep up with demographic trends and the explanations for them. Thus, every chapter of this twelfth edition has been revised for recency, relevancy, reliability, and readability.

- Chapter 1 (Introduction to Demography) updates the way in which demography connects the dots in the world, including a substantially revised essay on the "Mess in the Middle East."
- Chapter 2 (Global Population Trends) has been completely updated with the latest numbers on population rates of growth and geographic variability in demographic trends. The essay updates the prospects for countries currently confronting below-replacement level fertility.

- Chapter 3 (Demographic Perspectives) brings in the latest thinking on demographic theories, while at the same time emphasizing that the demographic transition is a whole suite of transitions, the discussion of which is really what the book is all about.
- Chapter 4 (Demographic Data) brings you the latest information about censuses and surveys throughout the world, with a special focus on the United States, Canada, and Mexico. There is also a revised section on spatial demography, along with a new essay on "The Demographics of Politics: Why the Census Matters."
- Chapter 5 (The Health and Mortality Transition) has all the latest numbers on disease and mortality, including a new discussion on trends in obesity that are showing up all over the world.
- Chapter 6 (The Fertility Transition) brings you new numbers and the latest thinking about how to accomplish low fertility, while at the same time avoiding fertility that some think is too low, with an emphasis in both instances on the role of women in society.
- Chapter 7 (The Migration Transition) updates the trends throughout the world in the movement of people between and within countries, with renewed emphases on the tragedies of being a refugee or slave in the modern world, and a full discussion of the ways in societies and migrants adapt to each other.
- Chapter 8 (The Age Transition) reviews the latest literature on the impact that changing age structures have on societies, and has updated data and projections of the older population in different countries in the world.
- Chapter 9 (The Urban Transition) includes the latest definitions of what constitutes an urban place, along with new data for cities, and a discussion of the sustainability of cities—one of the most pressing issues facing the future.
- Chapter 10 (The Family and Household Transition) has the most recent census data on the changing structures of families and households, especially in the United States, and has updated information about all of the elements of life chances that form a key section of this chapter.
- Chapter 11 (Population and Sustainability) has a new title and a new focus on sustainability instead of simply looking at environmental impacts of population growth and change. There is also an updated version of the very popular essay on the size of our ecological footprint, and a new focus on the question of whether population growth and our simultaneous quest for a higher standard of living are causing us to overshoot our carrying capacity.
- Chapter 12 (What Lies Ahead?) also has a new title and an increased emphasis on the likely future evolutions of population change in the world, along with a more tightly focused discussion of policy options available to us as we try to influence population trends and thus their impact on global change. The essay on the migration evolution in a local community also has a new surprise ending....

Special Features of the Book

To help increase your understanding of the basic concepts and issues of population studies, the book contains the following special features.

Short Essays Each chapter contains a short essay on a particular population concept, designed to help you better understand current demographic issues, such as Chapter 1's "Demographic Contributions to the 'Mess in the Middle East'" or Chapter 11's "How Big is Your Ecological Footprint." Each essay ends with two discussion questions to encourage you to think about the topic in greater depth.

Main Points A list of 10 main points appears at the end of each chapter, following the summary, to help you review chapter highlights.

Questions for Review A set of five questions are provided at the end of each chapter, designed to stimulate thinking and class discussion on topics covered in the chapter.

Websites of Interest At the end of each chapter, I have provided an annotated list of five websites that I have found to be particularly interesting and helpful to students.

Glossary A Glossary in the back of the book defines key population terms. These terms are in boldface type when introduced in the text to signal that they also appear in the Glossary.

Complete Bibliography This is a fully referenced book and all of the publications and data sources I have used are included in the Bibliography at the end of the book.

A Thorough Index To help you find what you need in the book, I have built as complete an index as possible, divided into a Subject Index, and a Geographic Index.

Ancillary Course Material

An Instructor's Manual and PowerPoint® lecture slides are available through the book's companion website, which you can access at login.cengage.com. Contact your sales representative for more information.

I regularly update my blog, providing resources for instructors and students: *http://weekspopulation.blogspot.com/.*

I also encourage you to download my Weeks Population app for your iPhone: **http://itunes.apple.com/app/weekspopulation/id491729979?mt=8**

Personal Acknowledgments

Like most authors, I have an intellectual lineage that I feel is worth tracing. In particular, I would like to acknowledge Kingsley Davis, whose standards as a teacher and scholar will always keep me reaching; Eduardo Arriaga; Judith Blake; Thomas Burch; Carlo Cipolla; Murray Gendell; Nathan Keyfitz; and Samuel Preston. Individually and collectively, they guided me in my quest to unravel the mysteries of how the world operates demographically. Thanks are due also to Steve Rutter, formerly of Wadsworth Publishing Company, who first suggested that I write this book. Special thanks go to John, Gregory, Jennifer, Suzanne, Amy, and Jim for teaching me the costs and benefits of children and children-in-law. They have instructed me, in their various ways, in the advantages of being first-born, the coziness of the middle child, the joys that immigration can bring to a family, the wonderful gifts (including Andrew, Sophie, Benjamin, Julia, Elizabeth, Kayla, and James) that daughters and daughters-in-law can bring, and the special place that a son-in-law has in a family.

However, the one person who is directly responsible for the fact that the first, second, third, fourth, fifth, updated fifth, sixth, seventh, eighth, ninth, tenth, eleventh and now the twelfth editions were written, and who deserves credit for the book's strengths, is my wife, Deanna. Her creativity, good judgment, and hard work in reviewing and editing the manuscript benefited virtually every page, and I have dedicated the book to her.

Other Acknowledgments

I would also like to thank the users of the earlier editions, including professors, their students (many of whom are now professors), and my own students, for their comments and suggestions. Many, many other people have helped since the first edition came out almost 40 years ago, and I am naturally very grateful for all of their assistance.

For this edition, I want to acknowledge that Dennis Hodgson of Fairfield University provided inspiration for parts of Chapter 3; Don Kerr of King's University College in Canada provided notes for Canada in Chapter 4; the essay in Chapter 7 was co-authored by Gregory B. Weeks, University of North Carolina at Charlotte; Marta Jankowska made important contributions to the essay in Chapter 12; and Sean Taugher updated the cartogram that provides the cover art. Thanks also for the many useful reviews of the eleventh edition that helped to inspire changes in this edition: Theodore Fuller, Virginia Tech, Blacksburg; Naomi Spence, Lehman College; Winfred Avogo, Illinois State University; Andrew Spivak, University of Nevada, Las Vegas; Susan Stewart, Iowa State University; and Philip Yang, Texas Woman's University.

CHAPTER 1
Introduction to Demography

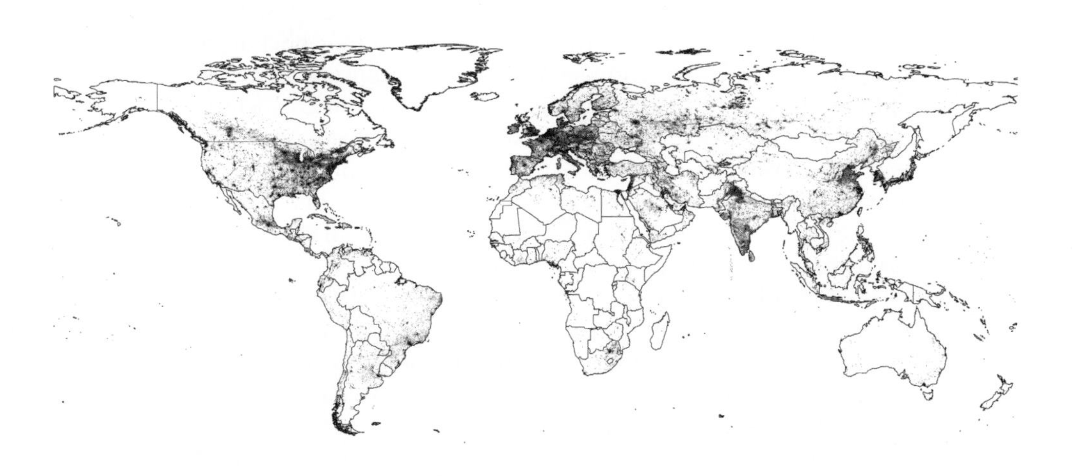

Figure 1.1 The World's Population Distribution Defined by Nighttime Lights
The nighttime light data are reversed on this map, so that the darker areas show where there are the most lights at night, suggestive of population density.

Source: Image and data processing by NOAA's National Geophysical Data Center, DMPS data collected by the US Air Force Weather Agency: http://www.gdc.noaa.gov/dmsp/.

WHAT IS DEMOGRAPHY?

HOW DOES DEMOGRAPHY CONNECT THE DOTS?

The Relationship of Population to Resources
- Food
- Water
- Energy
- Housing and Infrastructure
- Environmental Degradation

The Relationship of Population to Social and Political Dynamics
- Regional Conflict
- Globalization
- Immigration
- Riding the Age Wave
- The Relationship of Population to Rights of Women

How Is The Book Organized?

ESSAY: Demographic Contributions to the "Mess in the Middle East"

Population growth is an irresistible force. Indeed, every social, political, and economic problem facing the world today has demographic change as a root cause. What is more, I guarantee that it is a force that will increasingly affect you, personally, in ways both large and small throughout your life. Population change is not just something that happens to other people—it is taking place all around you, and you are making your own contribution to it.

The rise in life expectancy over the past two centuries, and most dramatically since the end of World War II, is the most important phenomenon in human history. More people living longer has produced unprecedented population growth and previously unthinkable transformations in human society. What is perhaps even more interesting to you, personally, is that this past is definitely prologue to your own future, as the world's population will continue to increase for the rest of your life. Though most of this growth will take place in developing countries (more specifically, in the *cities* of those countries), we will all experience the consequences: our own, as yet unthinkable, transformations.

Reports of declining birth rates in many parts of the world notwithstanding, it is a fact that the number of people added to the world each day is higher today than at any time in history. Moreover, we now live in a world crowded not only with people but also with contradictions. There are more highly educated people than ever before, yet also more illiterates; more rich people, but also more poor; more well-fed children, but also more hunger-ravaged babies whose images haunt us. We have better control over the environment than ever before, but we are damaging our living space in ways we are loath to imagine.

Our partial mastery of the environment is, indeed, key to understanding why the population is growing, because we have learned how to conquer more and more of the diseases that once routinely killed us. Although the rapid, dramatic drop in mortality all over the world is certainly one of humanity's greatest triumphs, we are finding that no good deed goes unpunished, even such an altruistic one as conquering (or at least delaying) death. Because the birth rate almost never goes down in tandem with the decline in the death rate, the result is rapid population growth. This relentless increase in numbers continues to fuel both environmental damage and social upheaval.

Demographic change isn't all bad news, of course, but population growth does make implacable demands on natural and societal resources. A baby born this year won't create much of an immediate stir outside her immediate family, but in a few years she will be eating more and needing clothes, an education, then a job and a place of her own. And, then, most likely, she will have babies of her own, and the cycle continues.

Understanding these and a wide range of related issues is the business of demography. Whether your concern with demography is personal or global or a combination, unraveling the "whys" of population growth and change will provide you with a better perspective on the world and how it works.

Demography is defined as the scientific study of human populations. But, really, demography is destiny. This book is an odyssey to understand the component parts of this powerful force and how they operate.

What Is Demography?

The term itself comes from the Greek root *demos*, which means people, and was coined in 1855 by Achille Guillard, who used it in the title of his book *Elements de Statistique Humaine ou Démographie Comparée*. Guillard defined demography as "the mathematical knowledge of populations, their general movements, and their physical, civil, intellectual and moral state" (Guillard 1855:xxvi). This is generally in tune with how we use the term today, in that modern demography is the study of the determinants and consequences of population change and is concerned with effectively everything that influences and can be influenced by:

- **population size** (how many people there are in a given place)
- **population growth or decline** (how the number of people in that place is changing over time)
- **population processes** (the levels and trends in fertility, mortality, and migration that are determining population size and change and that can be thought of as capturing life's three main moments: hatching, matching, and dispatching)
- **population spatial distribution** (where people are located and why)
- **population structure** (how many males and females there are of each age)
- **population characteristics** (what people are like in a given place, in terms of variables such as education, income, occupation, family and household relationships, immigrant and refugee status, and the many other characteristics that add up to who we are as individuals or groups).

It has been said that "the past is a foreign country; they do things differently there" (Hartley 1967:3). Population change and all that goes with it is an integral part of creating a present that seems foreign by comparison to the past, and it will create a future that will make today seem strange to those who look back on it several decades from now. Table 1.1 illustrates this idea, comparing population data for the United States in the year 1910 with that of the year 2010. To begin with, the top line of the table reminds us that in 1910 there were fewer than 2 billion people on the planet, whereas by 2010 there were nearly 7 billion (we hit that number in 2011). Although the U.S. population grew considerably during that century, from 92 million to 309 million, it did not keep pace with overall world population growth and so accounted for a slightly smaller fraction of the world's population in 2010 than it had in 1910 (more on this in the next chapter). Mortality levels in the U.S. dropped substantially over the century, leading to a truly amazing 29-year rise in life expectancy for females, from 52 in 1910 to 81 in 2010, with men lagging behind just a bit (the reasons for this are laid out in Chapter 5). Keep in mind that the life expectancy of 52 in 1910 was itself a big improvement over the 40 years people could expect to live in the middle of the nineteenth century.

Fertility also declined over the century between 1910 and 2010, although by world standards fertility in the United States in 1910 was already fairly low

Table 1.1 The Past Is a Foreign Country

	1910	2010
World population (billions)	1.8	6.9
U.S. population (millions)	92	309
U.S. percent of world total	5.1%	4.5%
Life expectancy (females)	52	81
Children per woman	3.5	1.9
Persons per household	4.4	2.6
% of U.S. population in California	3%	12%
Population of Buffalo, NY, compared to Los Angeles	Buffalo was more populous	LA was 15 times more populous
Immigrants from Italy (1900–1910); (2000–2010)	1.2 million	28,000
Immigrants from Mexico (1900–1910); (2000–2010)	123,000	1.7 million (legal immigrants)
% foreign-born	14.7%	12.9%
% urban	46%	81%
% of population under 15	32.1%	19.8%
% of population 65+	4.3%	13.0%
Passenger cars	450,000	190 million
% high school graduates among those 25 and older	~10%	87%

Source: Data for 1910 are from U.S. Census Bureau (1999); data for 2010 are from U.S. Census Bureau (2012) .U.S. Census Bureau: http://www.census.gov

(3.5 children per woman), having dropped from an estimated 7 children per woman at the beginning of the nineteenth century. Still, the drop from 3.5 to 1.9 clearly makes a huge difference in the composition of families, with average household size going down from 4.4 to 2.6 persons, and I discuss this more in Chapters 6 and 10.

Americans rearranged themselves spatially within the country over that span of time, and the considerable westward movement is exemplified by the increase in the fraction of the population living in California. It went from only 3 percent in 1910 to 12 percent in 2010. Consider that in 1910 Los Angeles had fewer people than Buffalo, New York; whereas by 2010 the Los Angeles metropolitan area was home to 15 times the number of people in Buffalo. In the latter part of the twentieth century, much of that growth in Los Angeles was fueled by immigrants from Mexico and Central America, but over the course of the century the composition of international immigrants had shifted substantially. In the decade preceding the 1910 census, there were about 123,000 Mexican immigrants to the United States, compared to 1.2 million Italian immigrants in the same time period. By contrast, in the decade leading up to the census in 2010, the numbers were essentially reversed, with 28,000 Italian immigrants and 1.7 million Mexican legal immigrants, in addition to a large number of undocumented immigrants. Yet, strange as it might seem in the current era, when there is so much talk about immigrants, the data in Table 1.1 show that the foreign-born population actually represented a greater fraction of the nation in 1910 than it did a century later. We'll explore the reasons for that in Chapter 7.

The past was young, with 32 percent under the age of 15 and only 4 percent aged 65 and older; whereas the present is older, with only 20 percent under 15 and 13 percent aged 65 and older (more on this in Chapter 8). The past was predominantly rural, and the present is predominantly urban (as I discuss in Chapter 9); the past was predominantly pedestrian (there were only 450,000 passenger cars in 1910), and the present is heavily dependent on the automobile (with more than 190 million passenger cars being driven around the country). In the past, people were considerably less well educated than today, with only about 10 percent of those in 1910 achieving a high school education, compared to 87 percent now. These trends are discussed more in Chapter 10.

The world of 1910 was very different from the world of 2010, and the demographics represent an important part of that difference. The future will be different, in its turn, partly because of demographic changes taking place even as you read this page. The study of demography is thus an integral part of understanding human society.

How Does Demography Connect the Dots?

It may sound presumptuous, even preposterous, to suggest that nearly everything is connected to demography, but it really is true. The demographic foundation of our lives is deep and broad. As you will see in this book, demography affects nearly every facet of your life in some way or another. Population change is one of the prime forces behind social and technological change all over the world. As population size and composition changes in an area—whether it be growth or decline—people have to adjust, and from those adjustments radiate innumerable alterations to the way society operates.

This is very different, however, from saying that demography determines everything. Demography is a force in the world that influences every improvement in human well-being that the world has witnessed over the past few hundred years. Children survive as never before, adults are healthier than ever before, women can limit their exposure to the health risks involved with pregnancy and still be nearly guaranteed that the one or two or three babies they have will thrive to adulthood. Having fewer pregnancies and babies in a world where most adults reach old age means that men and women have more "scope" in life: more time to develop their personal capacities and more time and incentive to build a future for themselves, their children, and everyone else. Longer lives and the societal need for less childbearing by women mean that the composition of families and households becomes more diverse. The changes taking place all over the world in family structure are not the result of a breakdown of social norms so much as they are the natural consequence of societies adapting to the demographic changes of people living longer with fewer children in a world where urban living and migration are vastly more common than ever before. These are all facets of demography affecting your life in important ways.

There is no guarantee, however, about how a society will react to demographic change. That is why it is impossible to be a demographic determinist. Demographic change does demand a societal response, but different societies will respond differently, sometimes for the better, sometimes not. Nonetheless, it turns out that population structures are sufficiently predictable that we can at least suggest the kinds of responses from which societies are going to have to choose. The population of the world is increasing by more than 200,000 people per day, as I will discuss in more detail in the next chapter, but this growth is much more intense in some areas of the world than in others. In those places where societies have been unable to cope adequately, especially with increasing numbers of younger people, the fairly predictable result has been social, economic, and political instability. At the other end of the spectrum, there is considerable angst in some of the richer countries in which very low fertility has pushed the population to the edge of a decline, if not already into decline.

Population change is obviously not the only source of trouble in the world, but its impact is often incendiary, igniting other dilemmas that face human society. Without knowledge of population dynamics, for example, we cannot fully understand why the world is globalizing at such a rapid pace, nor can we understand the roots of conflict from the Middle East to Southeast Asia; nor why there is a simultaneous acceptance of and a backlash against immigrants in the United States and Europe. And we cannot begin to imagine our future without taking into account the fact that the population of the world at the middle of this century is expected to include 2 to 3 billion more people than it does now, since the health of the planet depends upon being able to sustain a much larger number of people than are currently alive. Because so much that happens in your life will be influenced by the consequences of population change, it behooves you to understand the causes and mechanisms of those changes. Let's look at some examples.

The Relationship of Population to Resources

Food None of the basic resources required to expand food output—land, water, energy, fertilizer—can be considered abundant today. This especially impacts less developed countries with rapidly rising food demands and small energy reserves. Even now in sub-Saharan Africa, food production is not keeping pace with population growth, and this raises the fear that the world may have surpassed its ability to sustain even current levels of food production, much less meet the demands of the nearly 3 billion additional people who will be in line for a seat at the dinner table over the next few decades. And the problem is not just on land. The annual catch of wild fish leveled off in the 1990s and has been declining since then, with an increasing fraction of fish coming from farms harvesting the few species amenable to aquaculture.

Water An estimated one in three humans already face water scarcity, as demand for water increases faster than the available supply of fresh water. In theory, we can convert salt water (which is most of the water on the planet) into fresh water, but the process requires a lot of energy.

Energy Every person added to the world's population requires energy to prepare food, provide clothing and shelter, and to fuel economic life in general. Our rising standard of living is directly tied to our increasing use of energy, yet every increment in demand is another claim on those resources. We know that petroleum reserves are limited. Can we transition quickly enough to solar and/or wind energy to meet the needs of a growing population? No one knows. Will biofuels be the answer? Not likely, because they come from valuable crop land that we need for growing food.

Housing and Infrastructure All of the future population growth in the world is expected to show up in the cities, especially those in developing countries. The irony of growing more food is that it requires mechanization, rather than more laborers, so as the number of babies born in rural areas continues to exceed deaths, the "excess" population is forced to move to cities in hopes of finding a job there. This means building homes (which requires lumber, cement, and a lot of other resources) and providing urban infrastructure (water, sewerage, electricity, roads, telecommunications, etc.) for those 2 to 3 billion newcomers. This increasing "demographic overhead" is burdensome, particularly for those countries that already cannot adequately provide for their urban populations.

Environmental Degradation As the human population has increased, so has its potential for disrupting the earth's biosphere. The very same explosion in scientific knowledge that has allowed us to push death back to ever older ages, thus unleashing population growth, has also taught us how to convert the earth's natural resources into those things that comprise our higher standard of living. And it is not just that we are using up resources; waste accompanies use. The waste from fossil fuel use is carbon dioxide released into the atmosphere, generating the well-known effect on global climate change, evidenced perhaps most dramatically by the melting glaciers. But we are also damaging the hydrosphere (the world of water) by contaminating the fresh water supply, destroying coral reefs and fishing out the ocean, while also wreaking havoc on the lithosphere (the thin layer of the earth's crust upon which we live) by degrading the land with toxic waste and permitting top soil loss, desertification, and deforestation.

The task we will confront in the future is to maintain our standard of living while using many fewer resources per person. Keep in mind that international agencies such as the United Nations and the World Bank have suggested, through the Millennium Development Goals, that long-term sustainability of the planet requires that we lift all people out of poverty so that everyone can be a better steward of the planet. This is not going to be a simple project.

The Relationship of Population to Social and Political Dynamics

Regional Conflict Books and movies have been created to exploit the conflict that could be imagined if humans reached a point of diminishing resources. Back in 1967, even before the publication of Paul Ehrlich's *Population Bomb* (Ehrlich 1968),

Harry Harrison (1967) wrote a widely read book called *Make Room!Make Room!*, which in 1973 was made into a popular film called *Soylent Green*. This was a science fiction movie starring Charlton Heston and Edward G. Robinson in which they confront life way in the future in 2022 (oops, that's coming right up). This is a world suffering from overpopulation, depleted resources, poverty, dying oceans, and a hot climate due to the greenhouse effect—where much of the population survives on processed food rations, including "soylent green," which turns out to be "recycled" humans. A lot of similarly themed books and movies have come along since then, including one of my favorites, Dan Brown's best-selling book, *Inferno* (2013). In this thriller, a "brilliant lunatic" geneticist buys completely into a Malthusian "mathematical" view of the world (see Chapter 3 for a fuller discussion of Malthus) that humans will breed themselves into extinction, and so he unleashes a vector virus into the world to induce sterility.

Having thus far escaped these frightening scenarios, it is tempting to think that population growth has not really yet had much of an impact on civil society. That's because the real impact is harder to see, even if very real. It sort of creeps up on us one age group at a time, forcing families, communities, and then societies to adjust in some way or another. One reaction to population growth is to accept or even embrace the change and then seek positive solutions to the dilemmas presented by an increasingly larger (or smaller, for that matter) younger population (or older population)—you get the idea. Another reaction, of course, is to reject change. This is what the Taliban has been trying to do for decades in parts of Afghanistan and Pakistan—to prevent a society from modernizing by force and, in the process, keeping death rates higher than they might otherwise be (you will learn in Chapter 5 that Afghanistan has one of the highest rates of maternal mortality in the world, not to mention the deaths from the violence there), and maintaining women in an inferior status by withholding access to education, paid employment, health care, and the means of preventing pregnancy. The difficulty the Taliban (or any similar group) faces (besides active military intervention to stop them) is that it is very hard, if not impossible, to put the genie back in the bottle once people have been given access to a longer life and the freedoms that are inherently associated with that. Very few people in the world prefer to go back to the "traditional" life of harsh exposure to disease, oppression, and death.

The essay accompanying this chapter reviews the demographics of the Middle East and North Africa (MENA) region of the world, where we can see with special clarity the crucial role that demography has played for several decades now. For example, the migration of poor rural peasants to the cities of Iran, especially Tehran, contributed to the political revolution in that country back in 1978 by creating a pool of young, unemployed men who were ready recruits to the cause of overthrowing the existing government (Kazemi 1980; Lutz et al. 2010), and this pattern has been repeated throughout the region. It has been said that the "dogs of war" (with no disrespect meant to dogs) are young and male (Mesquida and Wiener 1999), and this description applies especially to the MENA region, where large fractions of the population are young, increasingly well educated, and frustrated by the lack of jobs, and where males are routinely accorded higher status than females.

The basic characteristic of a youth bulge is that a large fraction of the total population falls into the age range of approximately 15 to 29—old enough to be considered a young adult, but still young enough not to necessarily have settled into a job and family. We might think of this as an "incendiary" age group. If a country or region has too many people in this age cohort relative to the rest of the population, and they have a reason to be unhappy, trouble might be around the corner—it just needs some spark to ignite it (Weeks and Fugate 2012). Nearly a half-century ago, Moller (1968:246) argued that "in non-western nations, the outlook for young revolutionaries appears brightest where the poverty and insecurity of an underdeveloped but changing economy coincides with a high proportion of adolescents and young adults." This still seems like an accurate assessment of how demography and society interact to produce movements like Al-Qaeda.

Sub-Saharan Africa is another part of the world where population growth has been increasing faster than resources can be generated to support it—despite the devastation caused by HIV/AIDS—increasing the level of poverty and disease, and encouraging child labor, slavery, despair, and violent ethno-nationalist conflict (Wimmer et al. 2009). Throughout sub-Saharan Africa, the large number of children enmeshed in poverty and often orphaned because their parents have died of AIDS provides recruits for rebel armies waging warfare against one government or another. Those children who resist the army recruiters may find themselves sold into slavery, which is part of a larger global problem of child trafficking (International Labour Organization 2013). This kind of abuse of children is not caused by demographic trends, but the demographic structure of society contributes to the problem by creating a situation where disproportionate numbers of children are available to be exploited.

Globalization Regional conflict is one response to population growth, but a less violent, albeit still controversial, response has been globalization. Let me explain. Most broadly, globalization can be thought of as an increasing level of connectedness among and between people and places all over the world, although the term has taken on a more politically charged dimension since many people interpret it to mean a penetration of less developed nations by multinational companies from the more developed nations. This trend is promoted by the removal of trade barriers that protect local industries and by the integration of local and regional economies into a larger world arena. The pros and cons of this process invite heated debate, but an important, yet generally ignored, element of globalization is that it is closely related to the enormous increase in worldwide population growth that took place after the end of World War II.

Control over mortality, which has permitted the growth of population, occurred first in the countries of Europe and North America, and it was there that population first began to grow rapidly in the modern world, gaining steam in the late nineteenth and early twentieth centuries. However, after World War II, death control technology was spread globally, especially through the work of various UN agencies, funded by the governments of the richer countries. Since declines in mortality initially affect infants more than any other age group, there tends to be a somewhat delayed reaction in the realization of the effects of a mortality decline until those

DEMOGRAPHIC CONTRIBUTIONS TO THE "MESS IN THE MIDDLE EAST"

The Middle East and North Africa (MENA) region of the world refers to the following countries: Algeria, Bahrain, Djibouti, Egypt, Iran, Iraq, Israel, Jordan, Kuwait, Lebanon, Libya, Malta, Morocco, Oman, Qatar, Saudi Arabia, Syria, Tunisia, United Arab Emirates, West Bank and Gaza, and Yemen (see accompanying map and data table). The population of the MENA is mostly, although not entirely, Arab, and has been in demographic and political flux for a very long time.

The long-simmering tensions flared dramatically when a young Tunisian fruit vendor, Mohamed Bouazizi, was humiliated one time too many by a corrupt system and set himself on fire in protest in December 2010. His act of self-immolation in Tunisia ignited a wild fire that spread throughout the entire region. Thus began the Arab Spring or Arab Awakening that brought down not only the government of Tunisia, but also Libya and Egypt, and sparked a long civil war in Syria.

The politics underpinning the uprising stretch back decades, with the region especially roiled by the creation of the state of Israel in the late 1940s. At the end of World War I, the British took control of Palestine, which included the territory of what is now modern Israel and Jordan, from the remnants of the Ottoman Empire. As early as 1917, under the Balfour Declaration, the British had already agreed to help establish a Jewish national home in Palestine. Then, in the 1930s and 1940s, when European anti-Semitism encouraged the mass migration of Jews to Palestine, the resulting change in the demographics of the region led inexorably in the direction of a Jewish state. Not unexpectedly, this influx of European Jews was resisted, first by Palestinian Arabs and subsequently by virtually all Arab states.

In 1946, at the end of World War II, the modern state of Jordan was granted full independence, and Britain handed the decision about Palestine to the United Nations. Then, in 1947 the United Nations passed General Assembly Resolution 181, which ". . . provided for the creation of two states, one Arab and the other Jewish, in Palestine, and an international regime for Jerusalem. The Zionists approved of the plan, but the Arabs, having already rejected an earlier, more favorable (for them) partition offer from Britain, stood firm in their demand for sovereignty over Palestine in full" (Oren 2002:4). The stage was thus set for the continuing struggle for control of the region. The nascent state of Israel was immediately attacked by armies from all surrounding Arab nations but managed to prevail, and when hostilities ended in 1949 Israel had claimed more territory than originally allotted to it by the United Nations. Because as many as 750,000 of Palestine's Arabs (who came to be known simply as Palestinians) had fled the area when fighting broke out, the Jewish population emerged as the demographic majority. The Palestinian population was effectively cordoned into the Gaza Strip and the West Bank.

During the more than half century that the creation and continued existence of Israel has been a political issue on the world stage, the entire MENA region has been increasing dramatically in population size—always a powerful underlying force for change. The accompanying table shows that in 1950 MENA had a population of 81 million—almost exactly the same as the population of Japan in that year. But the estimated MENA population in 2015 will be 418 million—a 500 percent increase! By comparison, Japan had increased to only 126 million in 2015, only a 53 percent increase. The United Nations Population Division projects the MENA region to add nearly 200 million more people by 2050, to a whopping 604 million, while Japan is projected to decline down to 108 million.

As both populations and political tension explode, the region is also pushing hard against its environmental constraints—especially water. Thomas Friedman, writing in the *New York Times*, produced a very cogent analysis of the situation: "All these tensions over land, water and food are telling us something: The Arab awakening was driven not only by political and economic stresses, but, less visibly, by environmental, population and climate stresses as well. If we focus only on the former and not the latter, we will never be able to help stabilize these societies" (Friedman 2012).

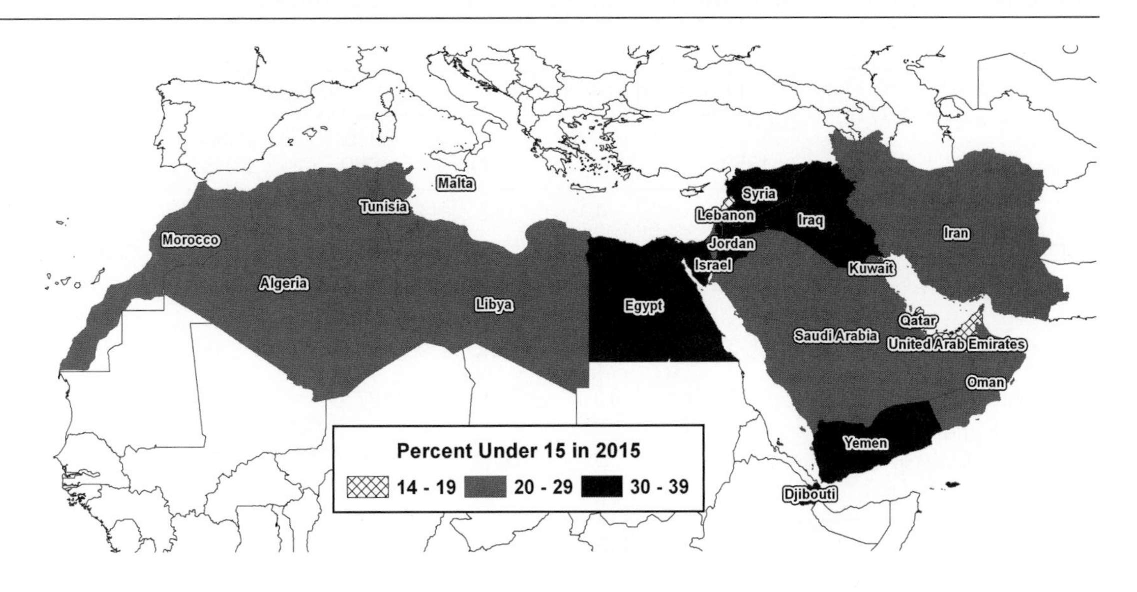

The Middle East and North Africa (MENA) Region
Source: See data in accompanying table.

Population Data for MENA

Country	Population (millions) in: 1950	2015	2050	Ratio of: 2015/1950	2050/2015	% < 15 in 2015
Algeria	9	41	55	4.6	1.3	28
Bahrain	0	1	2	14.0	1.3	22
Djibouti	0	1	1	15.0	1.3	34
Egypt	21	85	122	4.0	1.4	31
Iran	17	79	101	4.6	1.3	24
Iraq	6	36	71	6.0	2.0	39
Israel	1	8	12	8.0	1.5	28
Jordan	1	8	12	16.0	1.5	33
Kuwait	0	4	6	18.0	1.8	25
Lebanon	1	5	5	5.0	1.0	19
Libya	1	6	8	5.7	1.3	29
Malta	0	0	0	1.3	1.0	14
Morocco	9	34	43	3.8	1.3	28

(Continued)

DEMOGRAPHIC CONTRIBUTIONS TO THE "MESS IN THE MIDDLE EAST" (CONTINUED)

	Population (millions) in:			Ratio of:		
Country	1950	2015	2050	2015/ 1950	2050/ 2015	% < 15 in 2015
Oman	1	4	5	8.4	1.2	22
Qatar	0	2	3	80.0	1.3	14
Saudi Arabia	3	30	40	10.0	1.3	28
Syria	3	22	37	7.3	1.7	35
Tunisia	3	11	13	3.7	1.2	23
United Arab Emirates	0	10	16	137.1	1.6	16
West Bank and Gaza (Palestine)	1	5	9	5.0	2.0	39
Yemen	5	26	43	5.4	1.7	39
MENA Region	81	418	604	5.1	1.4	29
United States	*103*					*19*
Germany	*70*					*13*
Japan	*82*					*13*

Source: United Nations Population Division, World Population Prospects, 2012 Revision: http://www.un.org/en/development/desa/population/ (accessed 2013). *Note that projections are based on the medium fertility assumptions.*

The World Bank (2008) notes that MENA is the most water-scarce region in the world, and it is getting worse as the population grows.

The potential for political volatility can be seen in the age structure as shown in the accompanying table and map. In the region as a whole, nearly a third (29 percent) of the population is under the age of 15, but there is a spatial cluster of countries from Libya to the west all the way over to Iraq to the east in which 30 percent or more are under 15. As these young people emerge into adulthood, they will be confronting societies that likely will not have the resources to provide them with good jobs and a satisfactory life. That is a highly incendiary situation. Somewhat ominously, the two areas with the highest percent under 15 (39 percent) are Palestine (the combination of Gaza and the West Bank) and Iraq. Yemen, a country known to harbor terrorists, also has 39 percent of its population under the age of 15. For perspective, in the United States, 19 percent are under 15, and in both Germany and Japan only 13 percent are that young.

The bottom line is that this region is on course to continue its extreme rate of population growth in the face of serious environmental constraints. This suggests that regional stability may be a long way away, especially when you consider that four of the countries considered to be the world's most corrupt (Libya, Iraq, Syria, and Yemen) are in the region (Transparency International 2013).

Discussion Questions: (1) What do you think is the relationship between population growth in the MENA region and armed conflict? **(2)** How do you think the status of women in the Middle East might be influencing both demographic and political trends?

children who would otherwise have died reach an age where they must be educated, clothed, fed, and jobs and homes must be created for them on a scale never before imagined.

As huge new cohorts of young people have come of age and needed jobs in developing countries, their willingness to work for relatively low wages has not gone unnoticed by manufacturers in North America, Europe, and Japan. Nor have big companies failed to notice the growing number of potential consumers for products, especially those aimed at younger people, who represent the bulk of the population in developing countries. Given the demographics, it should not be surprising to us that jobs have moved to the developing countries and that younger consumers in those countries have been encouraged to spend their new wages on products that are popular with younger people in the richer countries, including music, fast food, cars, mobile phones, and electronic games.

Globalization of the labor market exists, in essence, because of the nature of world demographic trends. At the same time, the sheer volume of population growth in less developed countries is not a guarantee that jobs will head their way from richer countries. The likelihood goes up with two other demographically related factors: (1) declining fertility; and (2) increasing education. If fertility falls swiftly after mortality has gone down, the age structure goes through a transition in which there is a bulge of young adults ready to work, but they are burdened neither by a lot of dependent younger siblings nor yet by a lot of dependent older people. As I will discuss in more detail in Chapter 8, this "demographic dividend" can be used to good advantage, especially if a country (think China) has also spent societal resources educating its children so that the young people can readily step up to jobs that might be moved there from richer countries.

Immigration Globalization of the labor force has significantly broadened the ancient relationship between jobs and geography by bringing jobs to people in developing countries. For most of human history, a lack of jobs meant that young people moved to where the jobs were (or, at least, where they thought they were). That still happens. Even as some jobs are heading to developing countries, many young people in those countries are headed to the richer countries, facilitated by what I call the "demographic fit" between the young age structures of developing countries and the aging populations in richer countries.

The transition from higher to lower fertility in North America, Europe, and East Asia, as well as Australia and New Zealand, has created a situation in all of these parts of the world in which the younger population is declining as a fraction of the total, creating holes in the labor force and concerns about who will pay the taxes necessary to fund the pensions and health care needs of the elderly. For a variety of reasons that I will discuss in Chapter 6, women in the richer countries are choosing to have fewer children than are required to replace the population. On the other side of the coin are developing countries where, even if the birth rate is declining (as it is in most places), it hasn't declined as fast as the death rate, and so the young population keeps getting larger year after year. Supply meets demand in this demographic fit scenario, as low fertility countries take in migrants from higher fertility nations.

The United States has been the most accepting of all countries in the world in terms of absolute numbers of immigrants, including both legal and undocumented, with Mexico leading the list of countries from which immigrants to the U.S. come. Canada has been most welcoming of any country in the world on the basis of immigrants per resident population, with Asians being the largest group entering Canada (a pattern followed also in Australia).

Not to be overlooked, of course, is the fact that the countries sending migrants have their own demographic issues that complement those of the richer countries. For example, in Mexico, fertility decline for a long time had lagged behind the drop in mortality, and the resulting high rates of population growth made it impossible for the Mexican economy to generate enough jobs for each year's crop of new workers. The resulting underemployment in Mexico (people work, but there is not enough work to constitute a full-time job) naturally increased the attractiveness of migrating to where better jobs are. This happens especially to be the United States, not just because the United States is next door, but because low rates of population growth there have left many jobs open, particularly at the lower end of the economic ladder. These positions have provided foreign laborers with a higher standard of living than they could have in Mexico. The demographic dynamics have been shifting, though, and it seems likely that the demographic fit between the U.S. and Mexico is diminishing (Weeks and Weeks 2010). Fertility has been declining in Mexico, thus lowering the number of young people looking for work. This has helped the Mexican economy recover from the Great Recession of the first decade of this century, with the result that the supply of people thinking about heading to the U.S. has gone down. At the same time, the recent cohorts of immigrants to the U.S. have been having children, bolstering the number of younger people, and this, in combination with the slow recovery from the Great Recession, has lowered the demand for immigrant labor.

Because of limits on the number of legal immigrants admitted each year from specific countries in the world (see Chapter 7 for more details), a large fraction of those migrating from Mexico to the U.S. do so without documentation. However, since the terrorist attacks of September 11, 2001, undocumented immigrants have found it more difficult to enter the United States. As a consequence, many Latin American migrants have been going to Europe instead, both legally and illegally. The open-border policy within the European Union (EU) means that once people enter Europe, they are free to travel to any of the other EU countries in search of a job. Not surprisingly, Spain is the largest recipient of predominantly Spanish-speaking immigrants, but Switzerland and Italy also include growing communities of Latin Americans. There is a certain amount of symmetry, one might say, in the fact that the migration of Spaniards to the New World created "Latin America" from the mixing of Europeans with the indigenous population; now, five centuries later, the current is reversing.

There is, in fact, a bigger vacuum of laborers in Europe than in North America, because European birth rates have been declining for several decades and are now considerably lower than in the United States. There is thus the "sucking sound" of people from developing nations, notably former European colonies, filling the jobs in Europe that would otherwise go begging. The United Kingdom has large immigrant populations from India, Pakistan, and the Caribbean, whereas France has immigrants from Algeria and Senegal, Germany has immigrants from Turkey (not a former colony, but a sympathizer in both world wars), the Netherlands has immigrants from Indonesia, and Spain has immigrants from Morocco (along with those from Latin America). Europeans, however, are not necessarily in favor of this trend. Caldwell (2009) has documented the rise in anti-foreigner sentiment in Europe, aimed especially at Muslim immigrants, and politicians throughout Europe are increasingly being forced by voters to take a stand on immigration issues (Winter and Teitelbaum 2013).

Given the needs in European countries for laborers and the complementary surplus of laborers in developing countries, we can expect that immigration will quite literally change the face of Europe in your lifetime. The "demographic time bomb" of an aging European population (Kempe 2006) means that these countries could make good use of immigrants in place of the babies that aren't being born, but the problem is always that immigrants tend to be different. They may look different, have a different language, a different religion, and differ in their expectations about how society operates. Furthermore, since the immigrants tend to be young adults, they will wind up contributing disproportionately to the birth rate in their new countries, leading to a rapid and profound shift in the ethnic composition of the younger population. These differences create problems for all societies and create situations of backlash against immigrants.

American history is replete with stories of discrimination against immigrant groups for one or more generations until the children and grandchildren of immigrants finally are accepted as part of mainstream society. This process produces children who would not be recognizable to their ancestors and a society that is a foreign country relative to the past, as I discussed earlier in the chapter. Just as in the United States, European nations have highly visible anti-immigrant groups, but the immigrants have kept coming anyway because jobs were especially available in the run-up to the Great Recession, and are starting to come back again in the post-recession economic recovery.

By contrast, immigrants have not bolstered Japan's rapidly aging population because the level of anti-foreign sentiment is so high. The Japanese simply take it for granted that people from other countries will not become permanent members of Japanese society. This means that Japan has had fewer immigrant workers per person than North America or Europe, and it is not unreasonable to think that the Japanese economy has been moribund for many years now because it has not been invigorated by immigration.

Riding the Age Wave Grappling with uncertainty in the world requires more than guesswork, warned the late business guru Peter Drucker. It requires looking at "what has already happened that will create the future. The first place to look," said Drucker, "is in demographics" (quoted in Russell 1999:54). A key demographic with which societies must cope is the changing age structure. For example, if we go back only a few decades, we find that the demographics of the baby boom helped fuel inflation in the United States during the 1970s as government policies in that period were oriented toward creating new jobs for the swelling numbers of labor force entrants, directly contributing to inflation through government expenditures. This same bulge in the young adult male population also contributed to the ability of the United States to get as involved as it did in the Vietnam War—the "dogs of war" phenomenon mentioned earlier in connection with current issues in the Middle East.

The baby boom is still having an impact, but now the big question has become: How will the country finance the retirement and the health care needs of baby boomers as they age and retire? Most of the richest nations, but also China, are facing similar issues as declining fertility and increased longevity have contributed to the prospect of substantial increases in both the number and percentage of the older population (see Figure 1.2). As the older cohorts begin to squeeze national systems of social insurance, legislative action will be required to make long-run changes in the financing and benefit structure of these systems if they are to survive. As noted above and as I will discuss later in the book, immigration is one solution, but it comes with a lot of other costs attached. Changes will be made, of course, even if their exact shape is difficult to forecast. Delaying retirement is probably the easiest change to make, at least in the abstract. At the individual level, of course, few people want to make that choice to keep working for a few more years after spending their working life thinking that they were going to retire at a relatively young age. Increased self-reliance is another proposed solution, requiring people when younger to save for their own retirement through mandatory contributions to mutual funds

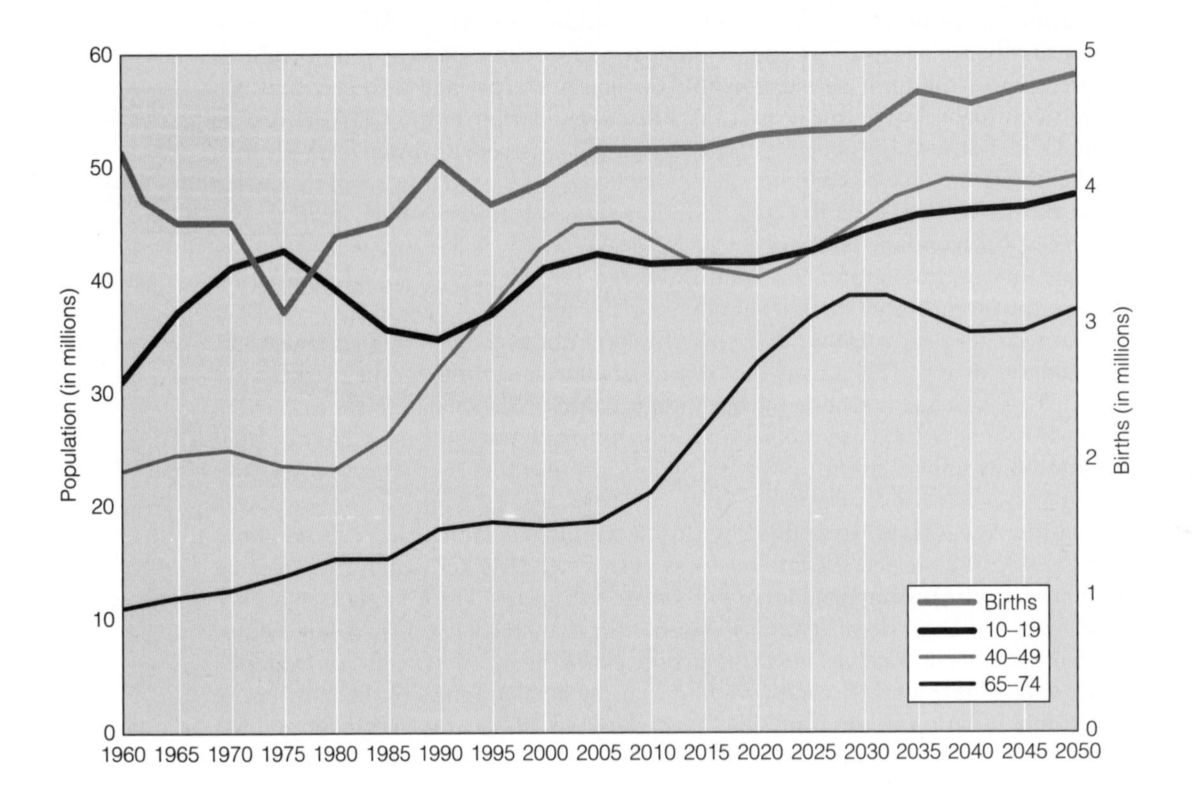

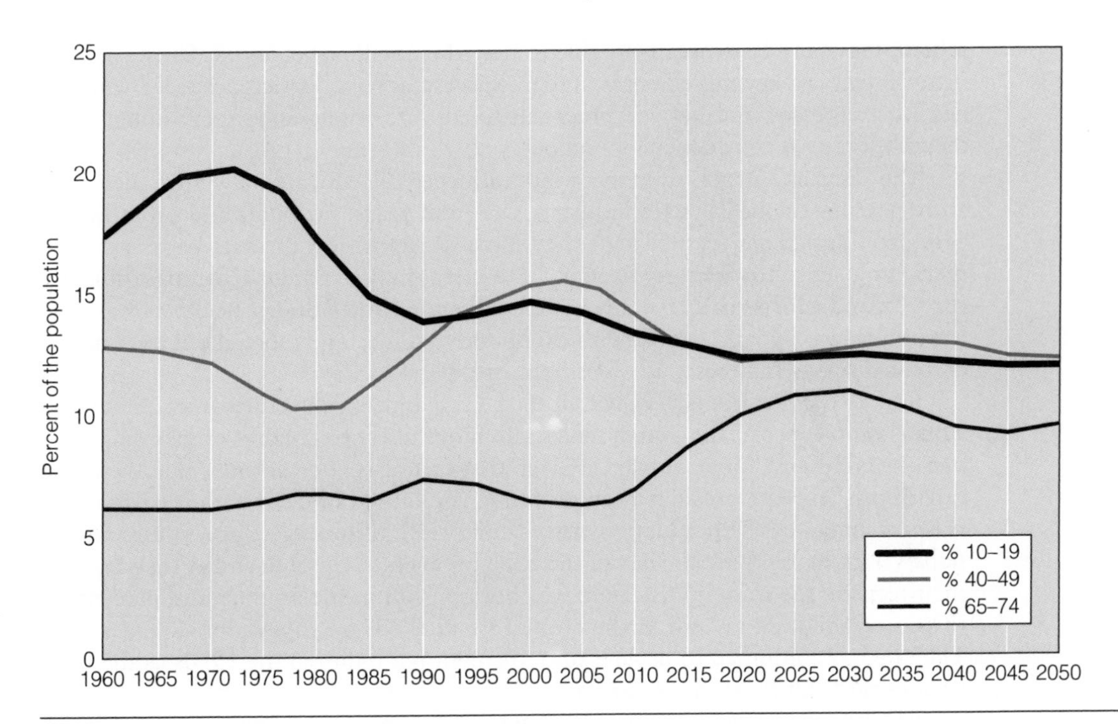

Figure 1.2 Riding the Age Wave: Number (top panel) and percentage (bottom panel) of selected age groups in the United States.

Source: Data for births 1960 to 1975 are from U.S. Census Bureau, 1996, Statistical Abstract of the United States (Washington, DC: Government Printing Office); Table 90; births for 1980 to 2010 are from the U.S. National Center for Health Statistics Vital Statistics Reports (various years); age data from 1960 to 2010 are from the United Nations Population Division; age and birth data from 2015 to 2050 are from the U.S. Census Bureau's National Population Projections (updated 2012): http://www.census.gov/population/projections/data/national/2012.html

and other investment instruments. It may also be, when the time comes, that taxes will be raised on younger people in order to bail out older people who, in fact, did not save enough for their retirement.

The changing age structure also has an obvious impact on the educational system. Public elementary and secondary school districts cannot readily recruit students or market their services to new prospects; they rise and fall on demographic currents that determine enrollment and the characteristics of students, such as English proficiency, that can affect resource demands. Of course, not every community experiences the wavy national trend shown in Figure 1.2; there is variation among and between individual school districts. All have a need for precise information about their particular area, because even within a district some geographic areas may be growing while others are diminishing in the number of school-aged children or children of one ethnic group or another. Demographic conditions can also affect a school district in ways that go beyond the numbers. Adult immigrants to the United States from Latin America, for example, tend to have low levels of education, so they may have relatively little experience with schools even in their native country, much less in the United States. When their children start attending school, they may be generally unable to help them with schoolwork. Since parental involvement is a key ingredient in student success, school districts are faced with the need to create new policies and programs to educate immigrant parents about what their children are experiencing in school.

The same age structure changes that influence the educational system also have an impact on the health care industry. Over the years, hospitals and other health care providers have learned that they have to reposition themselves in a classic marketing sense to meet the needs of a society that is changing demographically (Beckett and Morrison 2010). In countries like the United States, health care is now less about birthing and coping with childhood illnesses, and more about treating the chronic diseases that beset an older population.

Crime, like health, is closely tied to the age and sex structure of a community. Young people, especially young males, are more likely to commit crimes than anyone else (yet more "dogs of war"). Given that fact, it is not a surprise that the crime rate in the United States has been declining roughly in tandem with the decline in the percentage of the population that is comprised of teenagers and young adults. Babies born in 1964 are the last of the baby boomers in the U.S. and as they reached their teens in the mid-1970s, there was both a peak in the number and percentage of people aged 10-19 and in the crime rate in the U.S., especially violent crime. The rate has continued to decline since then, with a bump in the 1990s as the baby boomlet kids hit the crime-prone years.

The vast majority of people of any age, of course, are not criminals. But, almost all are consumers, and people at different ages have different needs and tastes for products and differing amounts of money to spend. Companies catering to the youngest age group have to keep track of the number of births (their potential market) as well as the characteristics of the parents and grandparents (who spend the money on behalf of the babies). The baby market has seen some wild fluctuations in recent decades in the United States, as you can see in Figure 1.2. The number of babies being born each year plummeted during the 1960s and 1970s, rebounded in the 1980s, peaked in 1990, and slacked off in the early 1990s before rebounding again. The U.S. Census Bureau projects the number of births to continue rising steadily for the foreseeable future. It is a dangerous business, however, to be lulled into believing that every company dealing with baby products will necessarily live or die on the peaks and troughs of birth cycles. In 2012, there were 3.9 million births in the United States, not far below the number at the height of the baby boom in 1958. Yet, in 2012 there were 1.6 million *first* births, compared to only 1.1 million in 1958. If you have hung around new parents and new grandparents, you know that people react differently to first births than to others. In particular, they open their pocketbooks wider to pay for cribs, buggies, strollers, diapers, and every conceivable type of baby toy designed to stimulate and "improve the quality" of the baby's life. Businesses that cater to these needs, then, are as sensitive to birth order as they are to the absolute volume of births.

The middle age group (represented in Figure 1.2 by people aged 40 to 49) was relatively unchanged in size since the 1960s and essentially unnoticed until the baby boom generation began moving into this category in the 1980s. Since then, serving them has become a new "boom" industry. Laser eye surgery surged, as did sales of walking shoes (running shoe sales slowed to a walk for boomers). When not walking, the aging baby boomers have been driving their luxury or near-luxury sport utility vehicles (and are now snapping up hybrid cars). It was the baby boom reaching middle age that helped to fatten the nonfat market, although younger people rather than baby boomers have led the movement toward vegetarian meals (Stahler 2012). They will probably carry those food preferences into their middle ages at a time when the number of people 40 to 49 will dip, as the baby boomers are replaced by the smaller cohort of Generation X.

The young-old population (ages 65 to 74) has been steadily increasing in numbers over time and, as you will learn in Chapter 8, has also become increasingly affluent. This segment of the population creates a market for a variety of things, from leisure travel to appliances with larger print, to door handles that are shaped to be used more easily by arthritic fingers. Perhaps most importantly, the aging of the population in North America has spurred the marketing of health services and products aimed at that age group, and the targeting of their wares to neighborhoods where people are aging in place (so-called naturally occurring retirement communities—NORC) (Morrison and Bryan 2010).

Johnson & Johnson provides a good example of a company that has kept its eye on the changing demographics not only of the United States but of the world in general. The company got its start in the 1880s when Robert Wood Johnson began selling sterile bandages and surgical products—innovations built on Lister's germ theory that helped to lower death rates in hospitals. Later on, during the years of the baby boom, Johnson & Johnson flourished by selling baby products. As the baby boom waned, the company continued to diversify its product line in a demographically relevant way, including acquiring ownership of both Ortho Pharmaceuticals (the largest U.S. manufacturer of contraceptives—helping to keep the birth rate low—and a large manufacturer of drugs to treat chronic diseases associated with aging) and Tylenol (one of the world's most popular pain relievers).

Basically, making sound investment decisions (as opposed to lucky ones) involves peering into the future, forecasting likely scenarios, and then acting on the basis of what seems likely to happen. After reading this book, you should have a good feel for the shape of things to come demographically. Most people do not, but those who do have an edge in life. A group of financial investors in the United Kingdom, for example, has established the Life and Longevity Markets Association in an attempt to spur the development of ways to make money from the pension funds into which an increasingly older population is pouring money. If people die sooner than expected, insurance companies lose money; whereas if they live longer than expected, the insurance companies reap a profit. The flip side of this is that if people live longer than expected, pension funds may be underfunded; whereas the pension funds profit if people die sooner, rather than later. You can see that people are betting one way or the other on your demographic future.

What else do the demographics suggest about future economic opportunities? The fact that 90 percent of the world's population growth in the foreseeable future will occur in the less developed nations is, as already noted, an important reason for the globalization of business and the internationalization of investment.

In 2012 two financial analysts in California put together a demographic-economic model of 176 countries of the world. Their conclusion was that age structures with a disproportionate share of people of working age are good for economic growth (economies with a demographic tailwind), and age structures with lots of kids or lots of older people are not so good (economies with a headwind). They summarize the situation as follows (Arnott and Chaves 2012:42):

> Children are not immediately helpful to GDP. They do not contribute to it, nor do they help stock and bond market returns in any meaningful way; their parents are likely disinvesting to pay their support. Young adults are the driving force in GDP growth; they are the sources of innovation and entrepreneurial spirit. But they are not yet investing; they are overspending against their future human capital. Middle-aged adults are the engine for capital market returns; they are in their prime for income, savings, and investments. And senior citizens contribute to neither GDP growth nor stock and bond market returns; they disinvest to buy goods and services that they no longer produce.

All is not lost, however, in those countries with lots of kids, because each one needs some kind of diaper. Procter & Gamble, maker of Pampers disposable diapers, has found a huge market out there. Babies grow up to be teenagers and young adults (trends that we will examine in detail throughout the book). From Malaysia to Argentina, young adults are buying iPads, cell phones, handheld electronic games, satellite dishes, and the perennial favorites, blue jeans and Coca-Cola. Companies selling in these markets are bound to make money.

International investors have been particularly intrigued by the world's two most populous countries, China and India. General Motors, Chrysler, and Ford all have invested in car manufacturing in China, as have Volkswagen and Peugeot Citroën from Europe. The problem, of course, is that a huge population does not necessarily mean a huge market if most people are poor. Starbucks serves coffee and Pizza Hut and McDonald's serve up fast food in China, but the average Chinese consumer cannot afford very many expensive goods. Enter Wal-Mart, which opened its first store in China in the mid-1990s and had 390 stores there as of 2013 (Walmart Stores 2013).

India, which is almost as populous, but is less well-off than China, does not yet allow full foreign ownership of retail businesses, except in very limited cases, but the so-called "consuming class" in India (those with at least some discretionary income, although it may be as low as $2 per day) is estimated to comprise about 300 million people (Mustafi 2013). This is about 25 percent of the population, yet it is a big market and thus represents an opportunity for some people to make money. Yum Brands, Inc., based in Louisville, Kentucky, which owns Pizza Hut, KFC, and Taco Bell (among other fast-food franchises), decided in 2010 that the growing young adult population (the youth bulge) in India represented a good market for Mexican food, and so they opened a Taco Bell in Bangalore focusing on the vegetarian aspects of their menu (Sharma 2013). Though we will return repeatedly to this paradox that many people (the "street") have a gut feeling that population growth is a good thing, we have no idea if we can sustain it, and if we can't, then what?

The Relationship of Population to Rights of Women There is probably no more important demographic issue than the rights of women. As I discuss in Chapter 5, women inherently have higher life expectancy than men, unless society intervenes to undermine that biological advantage. The other biological issue—reproduction—rears its head when society seeks to prevent women from controlling their own reproductive behavior, as I discuss in Chapter 6. In social terms, all evidence shows that men and women are equally able to be good or bad parents, equally able to become educated and succeed (or not) occupationally and economically, equally able to lead societies politically. Any group that oppresses women and suppresses their contributions will have a distinctively unfavorable demographic profile and will almost certainly suffer in terms of overall well-being. This theme will emerge regularly in subsequent chapters.

How Is the Book Organized?

In order to help you understand how the world works demographically in more detail, I have organized the book into four parts, each building on the previous one. This first chapter obviously is designed to introduce you to the field of population studies and illustrate why this is such an important topic. The second chapter reviews world population trends so that you have a good idea of what is happening in the world demographically, how we got to this point, and where we seem to be heading. The third chapter introduces you to the major perspectives or ways of thinking about population growth and change, and the fourth chapter reviews the sources of data that form the basis of our understanding of demographic trends.

In Part Two, "Population Processes," I discuss the three basic demographic processes whose transitions are transforming the world—the health and mortality transition (Chapter 5), the fertility transition (Chapter 6), and the migration transition (Chapter 7). Knowledge of these three population processes and transitions provides you with the foundation you need to understand why changes occur and what might be done about them.

Part Three, "Population Structure and Characteristics," is devoted to studying the interaction of the population processes and societal change that occur as fertility, mortality, and migration change. These include the age transition (Chapter 8), the urban transition (Chapter 9), and the family and household transition (Chapter 10). The fourth and final part of the book, "Using the Demographic Perspective," first explores the relationship between population and sustainability (Chapter 11): Can economic growth and development be sustained in the face of continued population growth? There are no simple answers, but we are faced with a future in which we will have to deal with the global and local consequences of a larger and constantly changing population. In Chapter 12, I review what lies ahead demographically and discuss the ways in which the global community is trying to cope politically with these changes as they alter the fabric of human society.

Summary and Conclusion

It is an often-repeated phrase that "demography is destiny," and the goal of this book is to help you to cope with the demographic part of your own destiny and that of your community, and to better understand the changes occurring all over the world. Demographic analysis helps you do this by seeking out both the causes and the consequences of population change. The absolute size of population change is very important, as is the rate of change, and of course, the direction (growth or decline).

The past 200 years have witnessed almost nonstop growth in most places in the world, but the rate is slowing down, even though we are continuing to add nearly 9,000 people to the world's total every hour of every day. You may not realize it, but everything happening around you is influenced by demographic events close to you as well as in faraway places. I refer not just to the big things like regional conflict, globalization, climate change, exhaustion of resources, and massive migration movements, but even to little things that affect you directly, like the kinds of stores that operate in your neighborhood, the goods that are stocked on your local supermarket shelf, the availability of a hospital emergency room, and the jobs aimed at college graduates in your community. Influential decision makers in government agencies, social and health organizations, and business firms now routinely base their actions at least partly on their assessment of the changing demographics of an area. So, both locally and globally, demographic forces are at work to change and challenge your future. The more you know about this, the better prepared you will be to deal with it (and perhaps even influence what the future will be). In the next chapter, I get you started on this by outlining the basic facts of the global demographic picture.

Main Points

1. Demography is concerned with everything that influences or can be influenced by population size, growth or decline, processes, spatial distribution, structure, and characteristics.
2. Almost everything in your life has demographic underpinnings that you should understand.
3. Demography is a force in the world that influences every improvement in human well-being that the world has witnessed over the past few hundred years.
4. The past was very different from the present in large part because of demographic changes taking place all over the globe; and the future will be different for the same reasons.
5. The cornerstones of population studies are the processes of mortality (a deadly subject), fertility (a well-conceived topic), and migration (a moving experience).
6. Demographic change demands that societies adjust, thus forcing social change, but different societies will respond differently to these challenges, sometimes for the better, sometimes for the worse.

7. Examples of global issues that have deep and important demographic components include the relationship of population to food, water, and energy resources, as well as housing and infrastructure, and environmental degradation.
8. Population is also connected to social and political dynamics such as regional conflict, often exacerbated by youth bulges, as well as globalization, the need for immigrants created by the phenomenon of "demographic fit" and then the backlash against those immigrants.
9. Changes in the age structure are the most obvious ways in which demography forces societal change and, at the same time, creates business opportunities—exemplified by the idea of "riding the age wave."
10. A key to all demographic trends in the world is the status of women.

Questions for Review

1. When did you first become aware of demography or population issues more broadly, and what were the things that initially seemed to be important to you?
2. Why is the idea that nearly everything is connected to demography, or the companion idea that demography is destiny, not the same as demographic determinism?
3. How do you think the politics of the Middle East and North Africa (MENA) will be influenced in the long term by the changing demographics of the region?
4. Discuss the relative advantages and disadvantages of a youth bulge for a society to deal with.
5. Because globalization has an underlying demographic component, how might that affect the investing patterns of someone who uses demography as one of their investment criteria?

Websites of Interest

Remember that websites are not as permanent as books and journals, so I cannot guarantee that each of the following websites still exists at the moment you are reading this. You may have to Google the name of the organization to find the current web address.

1. **http://www.census.gov**
   The website of the U.S. Census Bureau has many useful features, including U.S. and world population information and U.S. and world population clocks (where you can check the latest estimate of the total U.S. and world populations).
2. **http://www.prb.org**
   The Population Reference Bureau (PRB) in Washington, D.C., is a world leader in developing and distributing population information. The site includes regularly updated information about PRB's own activities, as well as links to a wide range of other population-related websites all over the world.

3. **http://www.un.org/en/development/desa/population/**
   The Population Division of the United Nations is the single most important producer of global demographic information, which can be accessed at this site. Closely related United Nations data can be accessed through **http://data.un.org.**
4. **http://www.poodwaddle.com/clocks/worldclock/**
   This website keeps track of a wide range of demographic data from various official sources and then produces estimates that are constantly being updated (thus, they are called "clocks") by extrapolation models.
5. **http://weekspopulation.blogspot.com/search/label/Introduction%20to%20Demography**
   Keep track of the latest news related to this chapter by visiting my WeeksPopulation website.

CHAPTER 2
Global Population Trends

Figure 2.1 Cartogram of Countries of the World by Population Size

Note: The map shows the size of each country of the world according to its population. Each square represents approximately 2 million persons.

Source: Prepared by John Weeks and Sean Taugher using data from the United Nations Population Division 2013, World Population Prospects 2012; data refer to medium projections for 2015.

WORLD POPULATION GROWTH
A Brief History
How Fast Is the World's Population Growing Now?
The Power of Doubling—How Fast Can Populations Grow?
Why Was Early Growth So Slow?
Why Are More Recent Increases So Rapid?
How Many People Have Ever Lived?
Redistribution of the World's Population through Migration
European Expansion
"South" To "North" Migration
The Urban Revolution

GEOGRAPHIC DISTRIBUTION OF THE WORLD'S POPULATION

GLOBAL VARIATION IN POPULATION SIZE AND GROWTH
North America
United States
Canada
Mexico and Central America
South America
Europe
North Africa and Western Asia
Sub-Saharan Africa
South and Southeast Asia
India, Pakistan, and Bangladesh
Indonesia and the Philippines
Vietnam and Thailand
Iran
East Asia
China
Japan
Oceania

ESSAY: Implosion or Invasion? The Choices Ahead for the Low-Fertility Countries

At this moment you are sharing the planet and vying for resources with more than 7 billion others, and before the year 2050 yet another 2 to 3 billion souls will have joined you at the world's table, as I noted in the previous chapter. This in and of itself is pretty impressive, but it becomes truly alarming when you realize what a huge leap up it is from the "only" 2.5 billion in residence as recently as 1950. This phenomenon is obviously without precedent, so we are sailing in uncharted territory. In order to deal intelligently with a future that will be shared with several billion more people than today, we have to understand why the populations of so many countries are still growing (and why others are not) and what happens to societies as their patterns of birth, death, or migration change. In this chapter, I trace the history of population growth in the world to give you a clue as to how we got ourselves into the current situation. Then I will take you on a brief guided tour through each of the world's major regions, highlighting current patterns of population size and rates of growth, with a special emphasis on the world's 10 most populous nations.

World Population Growth

A Brief History

Modern human beings have been around for at least 200,000 years (Cann and Wilson 2003; McHenry 2009), yet our presence on the earth was scarcely noticeable until very recently. For almost all of that time, humans were hunter-gatherers living a primitive existence marked by high fertility, high mortality, and at best only very slow population growth. Given the very difficult exigencies for survival in these early societies, it is no surprise that the population of the world on the eve of the **Agricultural Revolution** (also known as the **Neolithic Agrarian Revolution**) about 10,000 years ago is estimated to have been only 4 million people (see Figure 2.2).

Many people argue that the Agricultural Revolution occurred slowly but pervasively across the face of the earth precisely because the hunting-gathering populations were growing just enough to push the limit of the carrying capacity of their way of life (Boserup 1965; Cohen 1977; Harris and Ross 1987; Sanderson 1995). **Carrying capacity** refers to the number of people that can be supported indefinitely in an area given the available physical resources and the way in which people use those resources (Miller and Spoolman 2012). Since hunting and gathering use resources *extensively* rather than *intensively*, it was natural that over tens of thousands of years humans would move inexorably into the remote corners of the earth in search of sustenance. Eventually, people in most of those corners began to use the environment more intensively, leading to the more sedentary, agricultural way of life that has characterized most of human society for the past 10,000 years, starting first in what is now the Middle East, and then in what is now the eastern part of China.

The population began to grow more noticeably after the Agricultural Revolution. Between 8000 B.C. and 5000 B.C., about 333 people on average (births minus

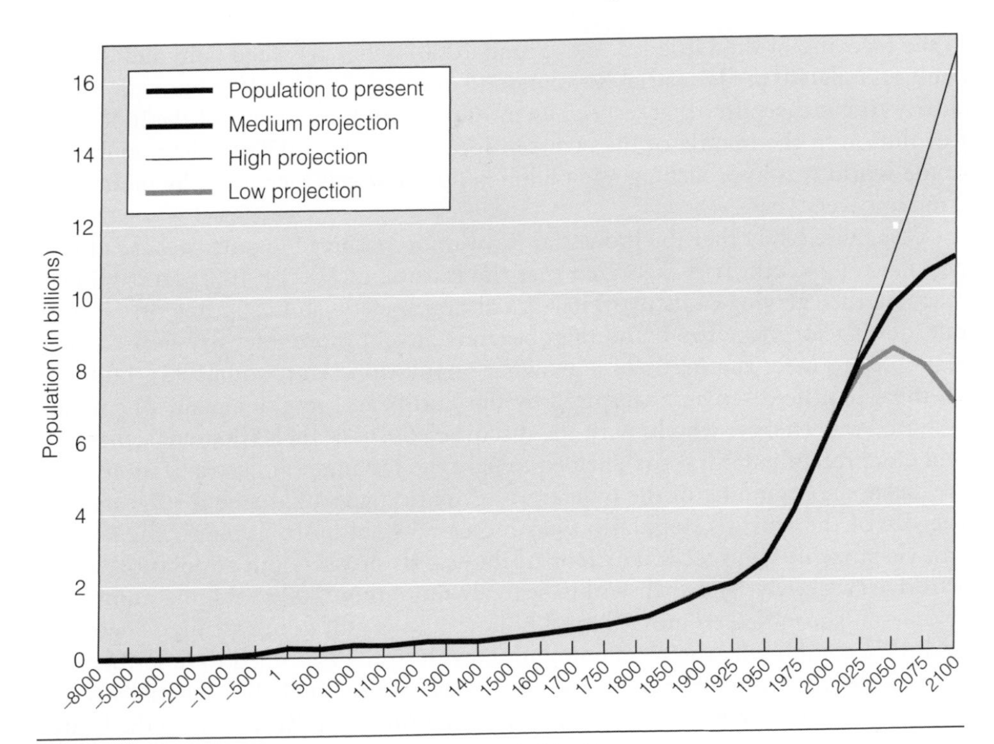

Figure 2.2 The World's Population Has Exploded in Size

Sources: The population data through 1940 are drawn from the U.S. Census Bureau, International Programs Center "Historical Estimates of the World Population," (http://www.census.gov/population/international/data/worldpop/table_history.php), accessed 2014. The numbers reflect the average of the estimates shown in that table. Population figures for 1950 through 2100 are from the United Nations Population Division, 2013, World Population Prospects: The 2012 Revision (http://esa.un.org/unpd/wpp/unpp/panel_population.htm), accessed 2014.

Note: Time is not to scale

deaths) were being added to the world's total population each year, but by 500 B.C., as major civilizations were being established in China, India, and Greece, the world was adding 100,000 people each year to the total. By the time of Christ (the Roman Period, A.D. 1) there may well have been more than 200 million people on the planet, increasing by nearly 300,000 each year.

There was some backsliding in the third through fifth centuries A.D., when increases in mortality, probably due to the plague, led to declining population size in the Mediterranean area as the Roman Empire collapsed, and in China as the Han Empire collapsed from a combination of flood, famine, and rebellion (McEvedy and Jones 1978). Population growth recovered its momentum only to be swatted down by yet another plague, the Black Death, that arrived in Europe in the middle of the fourteenth century and didn't leave until the middle of the seventeenth century (Cantor 2001).

After that, during the period from about 1650 to 1850, Europe as a whole experienced rather dramatic population growth as a result of the disappearance of the plague, the introduction of the potato from the Americas, and evolutionary (although not revolutionary) changes in agricultural practice—probably a response to the receding of the Little Ice Age (Fagan 2000)—that preceded (and almost certainly stimulated) the Industrial Revolution (Cohen 1995). The rate of growth began clearly to increase after that, especially in Europe, and on the eve of the Industrial Revolution in the middle of the eighteenth century (about 1750), the population of the world was approaching one billion people and was increasing by more than 2 million every year.

It is quite likely that the Industrial Revolution occurred in part because of this population growth. It is theorized that the Europe of 300 or 400 years ago was reaching the carrying capacity of its agricultural society, so Europeans first spread out looking for more room and then began to invent more intensive uses of their resources to meet the needs of a growing population (Harrison 1993), building on the scientific discoveries inspired by the European Enlightenment. The major resource was energy, which, with the discovery of fossil fuels (first coal, then oil, and more recently natural gas), helped to light the fire under industrialization.

Since the beginning of the Industrial Revolution approximately 250 years ago, the size of the world's population has increased even more dramatically, as you can visualize in Figure 2.2. For tens of thousands of years the population of the world grew slowly, and then within scarcely more than 200 years, the number of people mushroomed to more than 7 billion, and is still going strong. There can be little question why the term **population explosion** was coined to describe these historically recent demographic events. As you can see in Table 2.1, the world's population did not reach 1 billion until after the American Revolution—the United Nations fixes the year at 1804 (United Nations Population Division 1999)—but since then we have been adding each additional billion people at an accelerating pace. The 2 billion mark was hit in 1927, just before the Great Depression and

Table 2.1 The Billion People Progression

Year	Population in billions	Annual rate of growth	Annual increase in millions
1804	1	0.4	4
1927	2	1.1	22
1960	3	1.3	52
1974	4	2.0	75
1987	5	1.6	82
2000	6	1.4	77
2011	7	1.2	80
2024	8	0.9	73
2040	9	0.7	59
2061	10	0.4	38

Sources: The population data through 1940 are drawn from the U.S. Census Bureau, International Programs Center "Historical Estimates of the World Population," (http://www.census.gov/population/international/data/worldpop/table_history.php), accessed 2014. The numbers reflect the average of the estimates shown in that table. Population figures for 1950 through 2100 are from the United Nations Population Division, 2013, World Population Prospects: The 2012 Revision (http://esa.un.org/unpd/wpp/unpp/panel_population.htm), accessed 2014.

123 years after the first billion. In 1960, only 33 years later, came 3 billion; and 4 billion came along only 14 years after that, in 1974. We then hit 5 billion 13 years later, in 1987; we passed the 6 billion milestone 12 years later, in 1999; and in another 12 years, in 2011, we reached the seventh billion. The United Nations expects that we will reach 8 billion in 2023, 9 billion in 2040, 10 billion in 2061, and we could be very close to 11 billion by 2100—an incredible elevenfold increase in only three centuries (based on projections by the United Nations Population Division 2013).

I will discuss the methods of population projections in Chapter 8 and the implications of population projections in Chapter 12, but let me note that nearly everyone agrees that global population growth is likely to come to an end some time late in this century or early in the next century, even if we are not sure exactly when, nor exactly how many of us there will be when that time comes. The right side of Figure 2.2 shows the spread of options as calculated by the United Nations Population Division out to the end of this century. The middle projection, which the UN demographers think is the most likely, reaches 10.9 billion in 2100, as mentioned above, and assumes that the average number of births per woman in the world will eventually reach replacement level of just slightly more than two. The high projection of 16.6 billion in 2100 assumes that fertility levels decline but will not drop to replacement and so the world population keeps growing; whereas the low projection of 6.8 billion in 2100 assumes that fertility drops well below replacement, which of course would lead to eventual extinction if nothing changed.

How Fast Is the World's Population Growing Now?

The *rate* of population growth is obviously important (it is the "explosive" part), yet the *numbers* are what we actually cope with. In Figure 2.3, I have graphed the rate of population growth for the world over time. You can see that it peaked around 1970 and has been declining since then. This is good news, of course, but is tempered by the fact that as we continue to build on an ever larger base of people, the lower rates of growth are still producing very large absolute increases in the human population. When you have a base of more than 7 billion (our current population), the seemingly slow rate of growth of about 1.15 percent per year for the world as a whole still translates into the annual addition of 84 million people; whereas "only" 76 million were being added annually when the rate of growth peaked in 1970, and there were fewer than 30 million per year being added when the world was last growing at about 1 percent per year, at the end of World War II. Put another way, during the next 12 months, approximately 142 million babies will be born in the world, while 58 million people of all ages will die, resulting in a net addition of more than 84 million people. In the two seconds that it took you to read that sentence, nine babies were born while four people died, and so the world's population increased by five.

You can see a drop in the rate of population growth in the 1950–60 period. This was due to a terrible famine in China in 1959–60, which was produced by

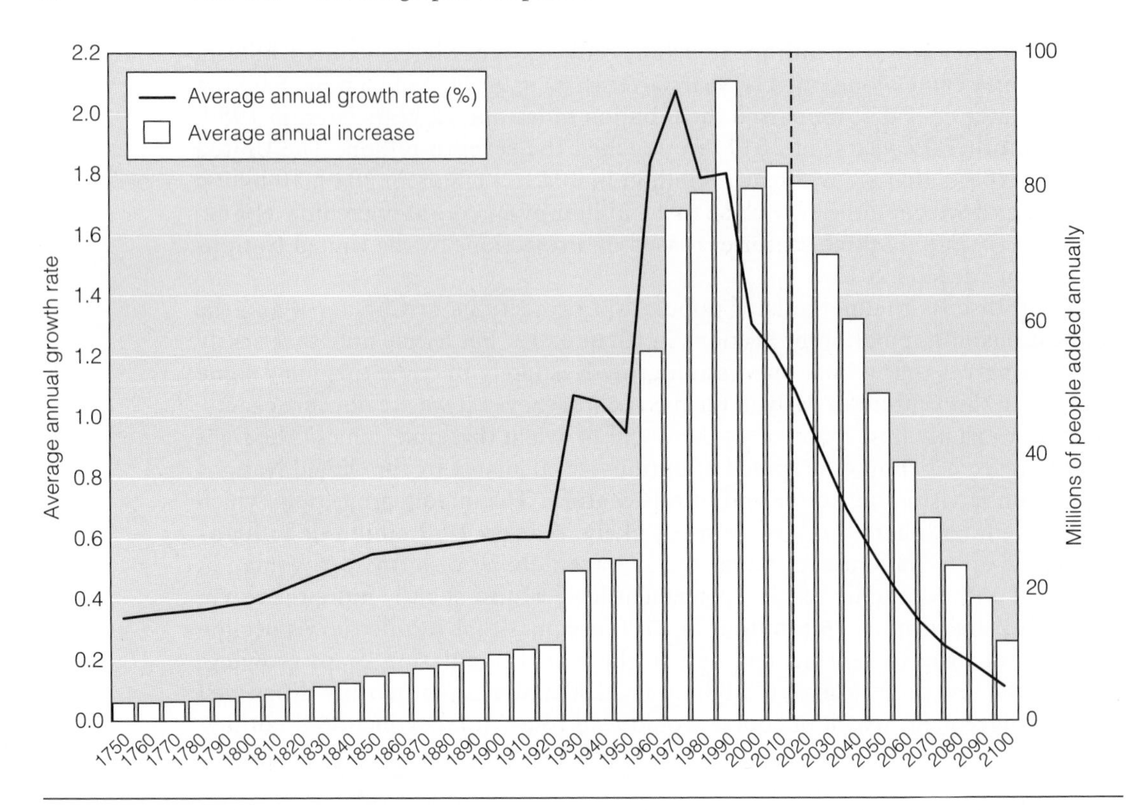

Figure 2.3 Tens of Millions of People Are Still Being Added to the World's Total Population Each Year, Despite the Drop in the Growth Rate

Sources: The population data through 1940 are drawn from the U.S. Census Bureau, International Programs Center "Historical Estimates of the World Population," (http://www.census.gov/population/international/data/worldpop/table_history.php), accessed 2014. The numbers reflect the average of the estimates shown in that table. Population figures for 1950 through 2100 are from the United Nations Population Division, 2013, World Population Prospects: The 2012 Revision (http://esa.un.org/unpd/wpp/unpp/panel_population.htm), accessed 2014.

Mao Zedong's Great Leap Forward program of 1958, in which the Chinese government "leapt forward" into industrialization by selling "surplus" grain to finance industrial growth. Unfortunately, the grain was not surplus, and the confiscation of food amounted to a self-imposed disaster that led to the deaths of 30 million Chinese in the following two years (1959 and 1960) (Becker 1997). Yet even though the Chinese famine was undoubtedly one of the largest disasters in human history, world population growth quickly rebounded, and the growth rate hit its record high shortly after that.

The Power of Doubling—How Fast Can Populations Grow?

Human populations, like all living things, have the capacity for exponential increase, which can be expressed nicely by the time it takes to double in population size. The incredible power of doubling is illustrated by the tale of the Persian chessboard. The story is told that the clever inventor of the game of chess was called in by the King of Persia to be rewarded for this marvelous new game. When asked what he would like his reward to be, his answer was that he was a humble man and deserved only a humble reward. Gesturing to the board of 64 squares that he had devised for the game, he asked that he be given a single grain of wheat on the first square, twice that on the second square, twice *that* on the third square, and so on, until each square had its complement of wheat. The king protested that this was far too modest a prize, but the inventor persisted and the king finally relented. When the Overseer of the Royal Granary began counting out the wheat, it started out small enough: 1, 2, 4, 8, 16, 32 . . . but by the time the 64th square was reached, the number was staggering—nearly 18.5 quintillion grains of wheat (about 75 billion metric tons!) (Sagan 1989). This, of course, exceeded the "carrying capacity" of the royal granary in the same way that successive doublings of the human population over the past 200 years threaten to exceed the carrying capacity of the planet.

Early on in human history it took several thousand years for the population to double to a size eventually reaching 14 million. From there it took a thousand years to nearly double to 27 million and another thousand to nearly double to 50 million, but less than 500 years to double from 50 to 100 million (about 500 years B.C.). Another thousand years later, in A.D. 500, it had doubled again. After yet another thousand years, in 1500, in the middle of the European Renaissance, the world's population had doubled again. About 400 years elapsed between the European Renaissance and the Industrial Revolution, and the world's population doubled in size during that time. But from 1750, it took only a little more than 100 years to double again, and the next doubling occurred in less than 100 years. The most recent doubling (from 3.5 to 7.0 billion) took only about 44 years.

Will we double again in the future? Probably not. Indeed we should hope not because we don't really know at this point how we will feed, clothe, educate, and find jobs for the 7 billion alive now, much less the additional 2 billion or more who are expected between now and later in this century. Once you realize how rapidly a population can grow, it is reasonable to wonder why early growth of the human population was so slow.

Why Was Early Growth So Slow?

The reason the population grew so slowly during the first 99 percent of human history was that death rates were very high. During the hunting-gathering phase of human history (hundreds of thousands of years), it is likely that life expectancy at birth averaged about 20 years (Petersen 1975; Livi-Bacci 2001). At this level of mortality, more than half of all children born will die before age five, and the average woman who survives to the reproductive years will have to bear nearly seven children in order to assure that two will survive to adulthood.

Research in the twentieth century on the last of the hunting-gathering populations in sub-Saharan Africa suggests that a premodern woman might have deliberately limited the number of children born by spacing them a few years apart to make it easier to nurse and carry her youngest child and to permit her to do her work (Dumond 1975). She may have accomplished this by abstinence, abortion, or possibly even infanticide (Howell 1979; Lee 1972). Similarly, sick and infirm members of society were at risk of abandonment once they were no longer able to fend for themselves. Not everyone agrees that there was any deliberate population control among early human populations, believing more simply that societies struggled to give birth to enough children to overcome the obstacle faced by the routinely high death rates among children (Caldwell and Caldwell 2003).

As humans settled into agricultural communities, population began to increase at a slightly higher rate than during the hunting-gathering era, and Bocquet-Appel (2008) has called this the **Neolithic Demographic Transition.** Initially it was thought that birth rates remained high but death rates declined slightly because of the more steady supply of food, and thus the population grew. However, archaeological evidence combined with studies of extant hunter-gatherer groups has offered a somewhat more complicated explanation for growth during this period (Spooner 1972). Fertility rates did, indeed, rise as new diets improved the ability of women to conceive and bear children (see Chapter 6). Also, it became easier to wean children from the breast earlier because of the greater availability of soft foods, which are easily eaten by babies. This would have shortened the birth intervals, and the birth rate could have risen on that account alone, and to a level higher than the death rate, thus promoting population growth. However, the sedentary life and the higher-density living associated with farming probably also *raised* death rates by creating sanitation problems and heightening exposure to communicable diseases. Nonetheless, growth rates increased even in the face of higher mortality as fertility rates rose to a level slightly higher than the death rate.

It should be kept in mind, of course, that only a small difference between birth and death rates is required to account for the slow growth achieved after the Agricultural Revolution. Between 8000 B.C. and 1750 A.D., the world was adding an average of only 67,000 people each year to the population. Right now, as you read this, that many people are being added every seven hours.

Why Are More Recent Increases So Rapid?

The acceleration in population growth after 1750 was due largely to the declines in the death rate that came about as part of the scientific revolution that accompanied the Industrial Revolution. First in Europe and North America and more recently in the rest of the world, death rates have decreased sooner and much more rapidly than have fertility rates. The result has been that many fewer people die than are born each year. In the more developed countries, declines in mortality at first were due to the effects of economic development and a rising standard of living—people were eating better, wearing warmer clothes, living in better houses, bathing more often, drinking cleaner water, and so on (McKeown 1976). These improvements in the human condition helped to lower exposure to disease and also to build up resistance to illness. Later, especially after 1900, much of the decline in mortality was due to improvements in public health and medical technology, including sanitation and especially vaccination against infectious diseases (Preston and Haines 1991).

Declines in the death rates first occurred only in those countries experiencing economic development. In each of these areas, primarily Europe and North America, fertility also began to decline within at least one or two generations after the death rate began its drop. However, since World War II, medical and public health technology has been available to virtually all countries of the world regardless of their level of economic development. In the less developed countries, although the risk of death has been lowered dramatically, birth rates have gone down less quickly, and the result is continuing population growth. As you can see in Figure 2.4, almost all the growth of the world's population is originating in less developed nations. I say "originating" because some of that growth then spills into the more developed countries through migration.

Between 2015 and 2050, the medium projections of the United Nations suggest that the world will add 2.2 billion people. Only 2 percent of this increase is expected to occur in the more developed nations. The less developed nations (excluding the least developed) will account for 59 percent of the increase, and the least developed will account for 39 percent. The least developed will be growing most quickly, increasing from 13 percent of the world's total in 2015 to 19 percent in 2050. On the other hand, the more developed nations will drop from 17 percent of the total in 2015 to less than 13 percent in 2050. These projections assume an actual decline in the size of the European population, where countries currently

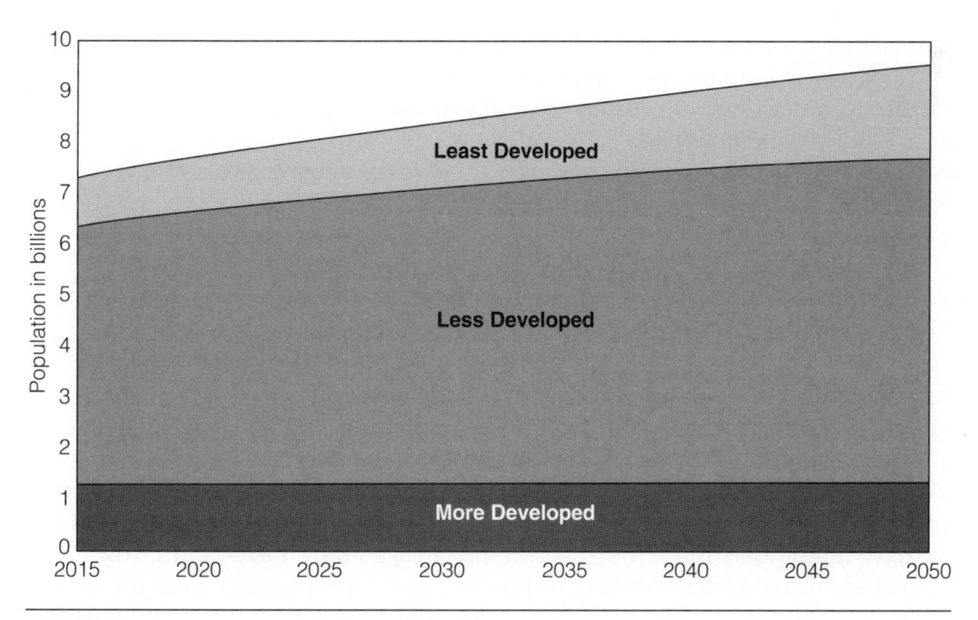

Figure 2.4 Less (and Least) Developed Regions Are the Sites of Future Population Growth

Source: Adapted by the author from Population Division of the Department of Economic and Social Affairs of the United Nations Secretariat, World Population Prospects: The 2012 Revision, http://esa.un.org/unpd/wpp/index.htm

have very low fertility alongside restrictive immigration laws. As I discuss in the essay accompanying this chapter, the aging of the European population and the threat of depopulation may well result in a loosening of the immigration laws.

How Many People Have Ever Lived?

The fact that we have gone from 1 billion to 7 billion in little more than 200 years has led some people to speculate that a majority of people ever born must surely still be alive. Let me burst that idea before it can take root in your mind. In fact, our current contribution to history's total represents only a relatively small fraction of all people who have ever lived. The most analytical of the estimates has been made by Nathan Keyfitz (Keyfitz 1966; Keyfitz and Caswell 2005), and I have used Keyfitz's formulas to estimate the number of people who have ever lived, assuming conservatively that we started with two people (call them "Adam and Eve" if you'd like) 200,000 years ago. The results of these calculations suggest that a total of 62.6 billion people have been born, of whom the 7.3 billion estimated to be alive in 2015 constitute 11.7 percent.

You can appreciate that the number of people ever born is influenced by the length of time you believe humans have been around, and by the estimate of the birth rate. There is no reasonable calculation, however, that generates a value much higher than 11.7, so we can safely assume that only a small fraction of humans ever born are now alive, although the percentage is constantly getting higher because of our ever larger population size. Furthermore, no matter how you calculate it, the dramatic drop in infant and childhood mortality over the last two centuries means that babies are now far more likely than ever to grow up to be adults. Thus, the adults alive today actually do represent a considerable fraction of all people who have ever lived to adulthood.

The vast increase in numbers is not the only important demographic change to occur in the past few hundred years. There has also been a massive redistribution of population.

Redistribution of the World's Population through Migration

As populations have grown unevenly in different areas of the world, the pressures or desires to migrate have also grown. This pattern is predictable enough that we label it the **migration transition** component of the overall demographic transition (which I will discuss in the next chapter). Migration streams generally flow from areas where there are too few jobs to areas where there is more opportunity. Thus we have migration from Latin America and Asia to the United States, from Asia to Canada, from Africa and Asia to Europe, and within Europe from the east to the west.

In earlier decades, the shortage of jobs generally occurred when the population grew dense in a particular region, and people then felt pressured to migrate to some other less populated area, much as high-pressure storm fronts move into low-pressure weather systems. This is precisely the pattern of migration that characterized the expansion of European populations into other parts of the world, as European farmers sought land in less densely settled areas. This phenomenon of European expansion is, of course, critically important because as Europeans moved around the world, they altered patterns of life, including their own, wherever they went.

European Expansion Beginning in the fourteenth century, migration out of Europe began gaining momentum, revolutionizing the entire human population in the process. With their gun (and germ)-laden sailboats, Europeans began to stake out the less-developed areas of the world in the fifteenth and sixteenth centuries—and this was only the beginning. Migration of Europeans to other parts of the world on a massive scale took hold in the nineteenth century, when the European nations began to industrialize and swell in numbers due to the decline in mortality. As Kingsley Davis has put it:

> Although the continent was already crowded, the death rate began to drop and the population began to expand rapidly. Simultaneous urbanization, new occupations, financial panics, and unrestrained competition gave rise to status instability on a scale never known before. Many a bruised or disappointed European was ready to seek his fortune abroad, particularly since the new lands, tamed by the pioneers, no longer seemed wild and remote but rather like paradises where one could own land and start a new life. The invention of the steamship (the first one crossed the Atlantic in 1827) made the decision less irrevocable. (Davis 1974:98)

Before the great expansion of European people and culture, Europeans represented about 18 percent of the world's population, with almost 90 percent of them living in Europe itself. By the 1930s, at the peak of European dominance in the world, people of European origin in Europe, North America, and Oceania accounted for 35 percent of the world's population (Durand 1967). By the beginning of the twenty-first century, the percentage had declined to 16, and it is projected to drop below 13 percent by the middle of the century (United Nations Population Division 2013). However, even that may be a bit of an exaggeration, since the rate of growth in North American and European countries is increasingly influenced by immigrants and births to immigrants from developing nations.

"South" to "North" Migration Since the 1930s, the outward expansion of Europeans has ceased. Until then, European populations had been growing more rapidly than the populations in Africa, Asia, and Latin America, but since World War II that trend has been reversed. The less developed areas now have by far the most rapidly growing populations, as we saw in Figure 2.4. It has been said that "population growth used to be a reward for doing well; now it's a scourge for doing badly" (Blake 1979). This change in the pattern of population has resulted in a shift in the direction of migration. For the past several decades there has been far more migration from less developed countries (the "South") to developed areas (the "North") than the reverse. Furthermore, since migrants are generally young adults of reproductive age, and since migrants from less developed areas generally have higher family size expectations than natives of the developed regions, their migration makes a disproportionate contribution over time to the overall population increase in the developed area to which they have migrated. As a result, the proportion of the population whose origin is one of the modern world's less developed nations tends to be on the rise in nearly every developed country. Within the United States, for example, non-Hispanic whites (the European-origin population) are no longer the majority in the state of California, and it is likely that the Hispanic-origin population (largely of Mexican ancestry) will represent the majority of Californians by the middle of this century since the majority of all births in California (as in all southwestern states) are now to Hispanic mothers. Note that I use "Hispanic" rather than the often-used "Latino" only because the former term is most often found in government statistics.

When Europeans migrated, they were generally filling up territory that had very few people, because they tended to be moving in on land used by hunter-gatherers who, as noted above, use land extensively rather than intensively. Those seemingly empty lands or frontiers have essentially disappeared today, and as a consequence migration into a country now results in more noticeable increases in population density. And, just as the migration of Europeans was typically greeted with violence from the indigenous population upon whose land they were encroaching, migrants today routinely meet prejudice, discrimination, and violence in the places to which they have moved. These days migrants are most likely to be moving to cities, because that's where the jobs are.

The Urban Revolution Until very recently in world history, almost everyone lived in basically rural areas. Large cities were few and far between. For example, Rome's population of 650,000 in A.D. 100 was probably the largest in the ancient world (Chandler and Fox 1974). It is estimated that as recently as 1800, less than 1 percent of the world's population lived in cities of 100,000 or more. Nearly half of all humans now live in cities of that size.

The redistribution of people from rural to urban areas occurred earliest and most markedly in the industrialized nations. For example, in 1800 about 10 percent of the English population lived in urban areas, primarily London. Now, 90 percent of the British live in cities. Similar patterns of urbanization have been experienced in other European countries, the United States, Canada, and Japan as they have industrialized. In the less developed areas of the world, urbanization was initially associated with a commercial response to industrialization in Europe, America, and Japan. In other words, in many areas where industrialization was not occurring, Europeans had established colonies or trade relationships. The principal economic activities in these areas were not industrial but commercial in nature, associated with buying and selling. The wealth acquired by people engaged in these activities naturally attracted attention and urban centers sprang up all over the world.

During the second half of the twentieth century, when the world began to urbanize in earnest, the underlying cause was the rapid growth of the rural population

(I discuss this in more detail in Chapter 9). The rural population in every less developed nation has outstripped the ability of the agricultural economy to absorb it. Paradoxically, in order to grow enough food for an increasing population, people have had to be replaced by machines in agriculture (as I will discuss in Chapter 11), and that has sent the redundant rural population off to the cities in search of work. Herein lie the roots of many of the problems confronting the world in the twenty-first century.

Geographic Distribution of the World's Population

The five largest countries in the world account for nearly half the world's population (an estimated 47 percent in 2015) but only 21 percent of the world's land surface. These countries include China, India, the United States, Indonesia, and Brazil, as you can see in Table 2.2. Rounding out the top 10 are Pakistan, Nigeria, Bangladesh, Russia, and Japan. Within these 10 most populous nations reside 58 percent of all people. You can see that you have to visit only the top 20 countries in order to shake hands with more than two out of very three people (69 percent) in the world. In doing so, you would travel across 40 percent of the earth's land surface. The rest of the population is spread out among 175 or so other countries that account for the remaining 60 percent of the earth's terrain.

If you set a goal to be as efficient as possible in maximizing the number of people you visit while minimizing the distance you travel, your best bet would be to schedule a trip to China and the Indian subcontinent. Four out of every ten people live in those two contiguous regions of Asia, and you can see how these areas stand out in the map of the world drawn with country size proportionate to population (see Figure 2.1, at the beginning of this chapter). Population growth in Asia is not a new story. In 1500, as Europeans were venturing beyond their shores, China and India (or more technically the Indian subcontinent, including the modern nations of India, Pakistan, and Bangladesh) were already the most populous places on earth, and all of Asia accounted for 53 percent of the world's 461 million people. Five centuries later, the population in Asian countries accounts for 61 percent of all the people on earth, although it is projected to drop back to 58 percent by the year 2050 because of their recent dramatic drop in fertility.

Sub-Saharan Africa, on the other hand, had about as many people as Europe did in 1500, comprising 17 percent of the world's population at that time. However, contact with Europeans tended to be deadly for Africans because of disease, violence, and especially slavery. It has been estimated that the export slave trade actually reversed African population growth from 1730 to 1850 (Manning and Griffith 1988). By the twentieth century, however, sub-Saharan Africa had rebounded in population size, comprising 13 percent of the total world population in 2015, and projected to be 22 percent of the total by 2050—even beyond where it had been in percentage terms in the year 1500, and with more than twice as many people as are projected to be in Europe in that year.

Table 2.2 The 20 Most Populous Countries in the World, 1950, 2015, and Projected to 2050

	1950			2015			2050		
Rank	Country	Population (in millions)	Area (000 sq miles)	Country	Population (in millions)	Area (000 sq miles)	Country	Population (in millions)	Area (000 sq miles)
1	China	563	3,601	China	1,402	3,601	India	1,620	1,148
2	India	370	1,148	India	1,282	1,148	China	1,385	3,601
3	Soviet Union	180	8,650	United States	325	3,536	Nigeria	404	352
4	United States	152	3,536	Indonesia	256	705	United States	401	3536
5	Japan	84	145	Brazil	204	3,265	Indonesia	321	705
6	Indonesia	83	705	Pakistan	188	298	Pakistan	271	298
7	Germany	68	135	Nigeria	183	212	Brazil	231	3,265
8	Brazil	53	3,265	Bangladesh	160	50	Bangladesh	202	50
9	United Kingdom	50	93	Russia	142	6,521	Ethiopia	188	386
10	Italy	47	114	Japan	127	705	Philippines	157	126
11	Bangladesh	46	50	Mexico	125	737	Mexico	156	737
12	France	42	212	Philippines	102	115	Congo (Kinshasa)	155	905
13	Pakistan	39	298	Ethiopia	99	386	Tanzania	129	366
14	Nigeria	32	352	Vietnam	93	126	Egypt	122	384
15	Mexico	28	737	Egypt	85	116	Russia	121	6,521
16	Spain	28	193	Germany	83	93	Japan	108	145
17	Vietnam	26	126	Iran	79	632	Uganda	104	93
18	Poland	25	118	Turkey	77	297	Vietnam	104	126
19	Egypt	21	384	Congo (Kinshasa)	71	905	Iran	101	632
20	Philippines	21	116	Thailand	67	197	Kenya	97	224
Top 20		1,958	23,977		5,083	23,449		6,413	23,600
World		2,556	57,900		7,325	57,900		9,551	57,900
% top 5		53%	29%		47%	21%		46%	17%
% top 10		65%	37%		58%	35%		53%	23%
% top 20		77%	41%		69%	40%		67%	41%

Source: Adapted by the author from Population Division of the Department of Economic and Social Affairs of the United Nations Secretariat, World Population Prospects: The 2012 Revision, http://esa.un.org/unpd/wpp/index.htm. Projections to 2050 are based on the medium fertility variant.

Global Variation in Population Size and Growth

World population is currently growing at a rate of 1.15 percent annually, implying a net addition of 84 million people per year, but there is a lot of variability underlying those global numbers. Germany and Russia, along with most countries in southern and eastern Europe, are expected to have fewer people in 2050 than in 2015. That is also the expectation for both China and Japan. Growth will take place largely in the less developed nations, as already shown in Figure 2.4, and in Figure 2.5 you can see where countries with the highest and lowest rates of growth are geographically. The most rapidly growing regions in the world tend to be in the mid-latitudes, and these are nations that are least developed economically—the "global south"; whereas the slowest growing are the richer nations, which tend to be more northerly and southerly (even though we label them as the "global north." It has not always been that way, however.

Before the Great Depression of the 1930s, the populations of Europe and, especially, North America were the most rapidly growing in the world. During the decade of the 1930s, growth rates declined in those two areas to match approximately those of most of the rest of the world, which was about 0.75 percent per year—a doubling time of 93 years. Since the end of World War II, the situation has changed again, and now Europe is on the verge of depopulation, while rapid growth is occurring in the less developed countries of Africa, Asia, and to lesser extent Latin America, which are now responsible for almost all of the world's population increase.

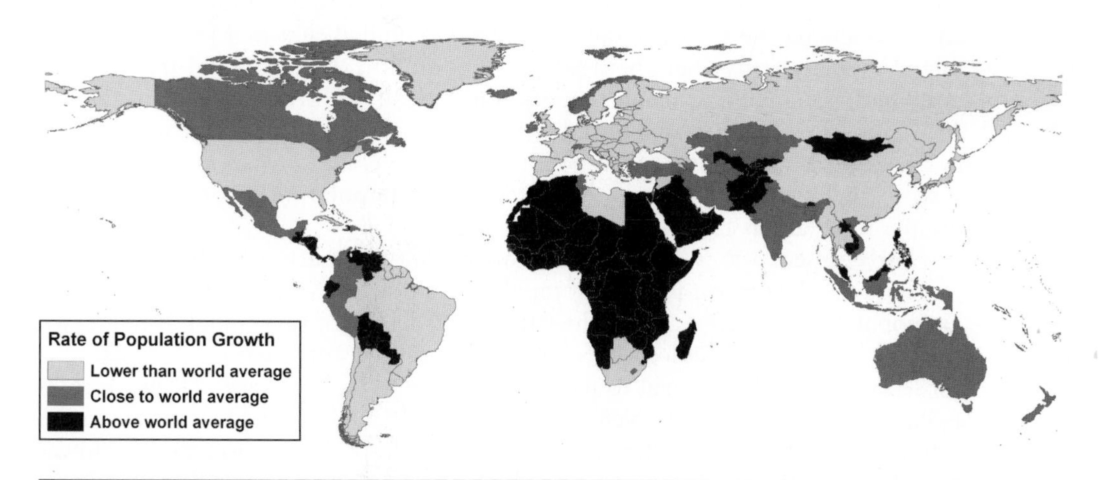

Figure 2.5 Rates of population growth are highest in the middle latitudes

Source: Adapted by the author from Population Division of the Department of Economic and Social Affairs of the United Nations Secretariat, World Population Prospects: The 2012 Revision, http://esa.un.org/unpd/wpp/index.htm.

Note: I defined close to world average as 0.95 to 1.35 percent per year; lower than world average was thus less than 0.95, and higher than world average was above 1.35.

Let's examine these trends in more detail, focusing particular attention on the 20 most populous nations, with a few other countries included to help illustrate the variability of demographic situations in which countries find themselves.

North America

The United States and Canada—North America—have a combined population of 361 million as of 2015, representing just under 5 percent of the world's total. The United States, with 325 million (third largest in the world), has 90 percent of North America's population, with Canada's 36 million people accounting for the remaining 10 percent. The demographic trajectories of the two countries are intertwined but are not identical.

United States It does not take a demographer to notice that the population of the United States has undergone a total transformation since John Cabot (an Italian hired by the British to search the new world) landed in Newfoundland in 1497 and claimed North America for the British. As was true throughout the western hemisphere, European guns and diseases rather quickly decimated the native American Indian population, making it easier to establish a new culture. Europeans had diseases and weapons that had never been seen by the indigenous populations, but the indigenous population had nothing that was new and dangerous to the Europeans, save perhaps for syphilis (Crosby 2004).

No one knows the size of the indigenous population in North America when the Europeans arrived, which of course leads to a lot of speculation (Mann 2011). After reviewing the evidence, Snipp (1989) concluded that in 1492 the number might have been anywhere between 2 and 5 million (keep in mind that Central and South America had much larger populations). We are more certain that by 1850, disease and warfare had reduced the native population to perhaps as few as 250,000, while the European population had increased to 25 million. Indeed, it was widely assumed that the American Indian population was on the verge of disappearing (Snipp 1989).

Early America was a model of demographic decimation for the indigenous population, while being a model of rapid population growth for the European-origin population. Yet even among the latter it was a land of substantial demographic contrasts. Among the colonies existing in the seventeenth century, for example, those in New England seem to have been characterized by very high birth rates (women had an average of seven to nine children) yet relatively low mortality rates (infant mortality rates in Plymouth Colony may have been as low or even lower than in some of today's less developed nations, apparently a result of the fairly good health of Americans even during that era) (Demos 1965; Wells 1971, 1982). Demos notes that "the popular impression today that colonial families were extremely large finds the strongest possible confirmation in the case of Plymouth. A sample of some ninety families, about whom there is fairly reliable information, suggests that there was an average of seven to eight children per family who actually grew to adulthood" (1965:270). In the southern colonies during the same time period, however, life was apparently much harsher, probably because the environment was more amenable to the spread of disease, including yellow fever and malaria. In the Chesapeake Bay colony of Charles Parish, higher mortality meant that few parents had more than two or three living children at the time of their death (Smith 1978).

Despite the regional diversity, the American population grew rather steadily during the seventeenth and eighteenth centuries, and though some of the increase in the number of Europeans in America was attributable to in-migration, the greater percentage actually was due to **natural increase** (the excess of births over deaths). The nation's first census, taken in 1790, shortly after the American Revolution, counted 3.9 million Americans, and although the population was increasing by nearly 120,000 a year, only about 3 percent of the increase was a result of immigration. With a crude birth rate of about 55 births per thousand population (comparable to the highest national birth rates in the world today) and a crude death rate of about 28 deaths per thousand (close to the highest in the world today), there were twice as many people being born each year as were dying. At this rate, the population was doubling in size every 25 years.

Though Americans may picture foreigners pouring in seeking freedom or fortune, it was not until the last third of the nineteenth century that migration became a substantial factor in American population growth. In fact, during the first half of the nineteenth century, immigration accounted for less than 5 percent of the population increase in each decade, whereas in every decade from the 1850s through the 1920s immigration accounted for at least 20 percent of the growth of population.

Throughout the late nineteenth and early twentieth centuries, the birth rate in the United States was falling. There is evidence that fertility among American Quakers began to be limited at about the time of the American Revolution (Wells 1971; Leasure 1989), and the rest of the nation was only a few decades behind their pace. By the 1930s, fertility actually dropped below replacement level (as I discuss more thoroughly in Chapter 6). Furthermore, since restrictions on immigration had all but halted the influx of foreigners during the Great Depression, Americans were facing the prospect of potential depopulation.

The early post–World War II era upset forecasts of population decline to be replaced by the realities of a population explosion. The result was the period from 1946 to 1964 generally known as the "baby boom" era. It was a time when the United States experienced a rapid rate of increase in population accomplished almost entirely by increases in fertility. The baby boom, in turn, was followed in the late 1960s and early 1970s by a "baby bust" (now widely known as Generation X, as I noted in the previous chapter). The birth rate bottomed out in 1976 and has been higher ever since, although not by very much (Martin et al. 2013). An echo of the baby boom was experienced as the "baby boomlet" of the 1980s, due largely to an increase in births by children of the baby boom, rather than an increase in the birth rate per woman.

Since the 1980s, fertility in the United States has remained at or just below the replacement level. Despite this low fertility, however, population growth has continued to be the order of the day largely because in the 1960s and then again in the 1990s adjustments of the nation's immigration laws opened the doors wider. Indeed, the 1 million immigrants (legal and estimated undocumented) being added each year account for nearly 40 percent of the annual increase in population. More importantly, from a demographic perspective, immigrants are primarily people of reproductive age and they are having children at a rate that is above replacement level. Indeed, variations in fertility levels in the United States are increasingly determined by fertility differences among the various racial and ethnic groups.

Canada The French were the first Europeans to settle the area that has become Canada, but in 1763 the French government ceded control of the region to the British, and a century later the British North America Act of 1867 united all of the provinces of Canada into the Dominion of Canada, and every census since then has asked about "origins" as a way of keeping track of the numerical balance between the historically rival French-speaking and English-speaking groups (Boyd et al. 2000).

In the seventeenth and eighteenth centuries, the high fertility of French speakers in Canada was legendary and they maintained higher-than-average levels of fertility until the 1960s (Beaujot 1978), probably due to the strong influence of the Catholic Church in Québec (McQuillan 2004). In the rest of Canada, fertility began to drop in the nineteenth century and, as in the United States, reached very low levels in the 1930s before rebounding after World War II in a baby boom that was similar in its impact on Canadian society to that experienced in the United States. This boom was similarly followed by a baby bust and then a small echo of the baby boom (Foot 1996). Canada (including the province of Québec) now has a fertility level that is well below replacement (1.6 children per woman).

Just as fertility is lower in Canada than in the United States, so is mortality, with life expectancy in Canada about two years longer than in the United States. In both of these respects the demographic profile of Canada is more like that of Europe than of the United States. However, when it comes to immigration, Canada more closely reflects the Northern American history of being a receiving ground for people from other nations. Despite its lower fertility, Canada's overall rate of population growth exceeds that of the United States because it accepts more immigrants per person than the United States does (though the U. S. accepts a higher total number).

Mexico and Central America

Mexico and the countries of Central America have also been growing since the end of World War II as a result of rapidly dropping death rates and birth rates that have only more recently begun to drop. With a population in 2015 of 125 million (eleventh most populous in the world), Mexico accounts for about three-fourths of the population of the region, with the remainder distributed among (in order of size) Guatemala, Honduras, El Salvador, Nicaragua, Costa Rica, Panama, and Belize. The combined regional population of 170 million as estimated for the year 2015 is a little more than 2 percent of the world's total.

Indigenous populations in Mexico and Central America had developed more advanced agricultural societies than had those in North America at the time of European contact. The Aztec civilization in central Mexico and the remnants of the Mayan civilization farther south centered near Guatemala encompassed many millions more people than lived on the northern side of what is now the United States–Mexico border. This fact, combined with the Spanish goal of extracting resources (a polite term for plundering) from the New World rather than colonizing it, produced a very different demographic legacy from what we find in Canada and the United States.

Mexico was the site of a series of agricultural civilizations as far back as 2,500 years before the invasion by the Spanish in 1519. Within a relatively short time after Europeans arrived, however, the population of several million was cut by as much as 80 percent due to disease and violence. This population collapse (a true implosion) was precipitated by contact with European diseases, but it reflects the fact that mortality was already very high before the arrival of the Europeans, and it did not take much to upset the demographic balance (Alchon 1997). By the beginning of the twentieth century, life expectancy was still very low in Mexico, less than 30 years (Morelos 1994), and fertility was very high in response to the high death rate. However, since the 1930s the death rate has dropped dramatically, and life expectancy in Mexico is now 79 years for women, six years above the world average and only two years less than in the United States.

For several decades, this decline in mortality was not accompanied by a change in the birth rate, and the result was a massive explosion in the size of the Mexican population. In 1920, before the death rate began to drop, there were 14 million people in Mexico (Mier y Terán 1991). By 1950 that had nearly doubled to 26 million, and by 1970, it had nearly doubled again to 49 million. In the 1970s, the birth rate finally began to decline in Mexico, encouraged by a change in government policy that began promoting small families and the provision of family planning. Mexican women had been bearing an average of six children each for many decades (if not centuries), but by 2015 this figure had dropped to 2.2 children per woman for the country, and in the capital of Mexico City fertility has dropped below replacement level to 1.8 children per woman (INEGI 2013). Nonetheless, until very recently the massive buildup of young people strained every ounce of the Mexican economy, encouraging outmigration, especially to the United States.

The other countries of Central America have experienced similar patterns of rapidly declining mortality, leading to population growth and its attendant pressures for migration to other countries where the opportunities might be better. Not every country in the region has experienced the same fertility decline as Mexico, however. In particular, Guatemala, Nicaragua, and Honduras are countries in which a high proportion of the population is indigenous (rather than being of mixed native and European origin), and they have birth rates that are well above the world average. Indeed, the average woman in Guatemala is having 3.9 children. However, Costa Rica has a more European pattern, with fertility that is below replacement, life expectancy that is the same as in the United States, and the need to import workers from its neighboring countries.

South America

The 415 million inhabitants of South America as of 2015 represent about 6 percent of the world's total population, with Brazil alone accounting for half of that. Its population of 204 million makes it the fifth most populous nation in the world. The modern history of Brazil began when Portuguese explorers found an indigenous hunter-gatherer population in that region and tried to enslave them to work on plantations. These attempts were unsuccessful, and the Portuguese wound up populating the colony largely with African slaves. The 4 million slaves taken to Brazil represent more than one-third of all slaves transported from Africa to the western hemisphere between the sixteenth and nineteenth centuries (Thomas 1997). The Napoleonic Wars in Europe in the early part of the nineteenth century allowed Brazil, like most Latin American countries, to gain independence from Europe, and the economic development that followed ultimately led to substantial migration into Brazil from Europe during the latter part of the nineteenth century. The result is a society that is now about half European-origin and half African-origin or mixed race. Brazil also attracted Japanese immigrants early in the twentieth century and again after World War II.

Between the 1960s and the 1990s, Brazil experienced a reduction in fertility described as "nothing short of spectacular" (Martine 1996:47). In 1960, the average Brazilian woman was giving birth to more than 6 children, but it had dropped to 2.5 by 1995 and it has since declined even further to 1.8—below replacement level. For many years, the influence of the Catholic Church was strong enough to cause the government to forbid the dissemination of contraceptive information or devices, but economic development beginning in the1960s seems clearly to have encouraged a decline in fertility, even if that drop was spatially uneven—starting first in the south and southeast and then spreading north and northeast from there (Schmertmann et al. 2008). The infant mortality rate of 21 deaths per 1,000 live births in Brazil is half the world average, and female life expectancy of 78 years is well above the world average.

The other countries of South America can be loosely divided into those three most southerly nations (including Argentina, Chile, and Uruguay) in which the populations are predominantly of European origin, and the remaining countries with higher proportions of indigenous and mixed (mestizo) populations. The former tend to have lower fertility and higher life expectancy—levels very similar to Brazil's. The other countries tend to have higher fertility, somewhat higher mortality, and higher rates of population growth.

Europe

The combined population of western, southern, northern (including Scandinavia), central, and eastern Europe (including Russia) is about 743 million, or about 11 percent of the world's total. Russia is currently the most populous, accounting for 19 percent of Europe's population. With 142 million, it is the ninth most populous country in the world. The next most populous countries in Europe are, in order, Germany (sixteenth most populous in the world), the United Kingdom, France, and Italy and they, along with Russia, comprise more than half (56 percent) of all Europeans.

Europe as a region is on the verge of depopulating. This is largely because its two largest nations, Russia and Germany, currently have more deaths than births and neither country is taking in enough immigrants to compensate for that fact. But the threat of depopulation actually extends to every country in eastern and southern Europe. No country in that part of Europe is projected to have more people in 2050 than they currently have, whereas nearly every one of the northern and western European countries is expected to have the same number or slightly more people by mid-century.

It is not a coincidence that German demographics look more like eastern than western Europe. When East Germany was reunited with West Germany, the combined Germany inherited the East's dismal demographics and that largely explains why Germany teeters on depopulation. Russia's situation is especially noteworthy because depopulation is not just a result of below replacement fertility. Until very recently, life expectancy for males was actually declining, signaling major societal stresses. In fact, researchers have argued that the breakup of the Soviet Union was foreshadowed by its rise in death rates (Feshbach and Friendly 1992; Shkolnikov et al. 1996). The birth rate was already low in Russia before the breakup, and since then the average number of children being born per woman has remained well below replacement level.

The rest of Europe has experienced very low birth rates without the drop in life expectancy that has plagued Russia. Where population growth is occurring, such as in France, the United Kingdom, and Ireland, it is largely attributable to the immigration of people from less developed nations as well as from eastern Europe.

It should not be a surprise that fertility and mortality are both low in Europe, since that is the part of the world where mortality first began its worldwide decline approximately 250 years ago and where fertility began *its* worldwide decline about 150 years ago. What is surprising, however, is how low the birth rate has fallen. It is especially low in the Mediterranean countries of Italy and Spain, where fertility has dropped well below replacement level—in predominantly Catholic societies where fertility for most of history has been higher than in the rest of Europe. Sweden and the other Scandinavian countries have emerged with fertility rates that are now among the highest in Europe, although still below the replacement level, and as I discuss in both Chapters 6 and 10, it is likely that an improvement in the status of women may be required to push fertility levels in Europe back up to the replacement level. Throughout eastern and southern Europe women have been given the opportunity for education and a career, but traditional attitudes remain in terms of their domestic role. Combining a career with family-building is generally frowned upon and, in response, women have tended to choose a career and either a small family or no family at all.

I have already mentioned that a major consequence of the low birth rate is an aging of the population that has left Europe with too few young people to take jobs and pay taxes. Into this void have swept millions of immigrants, many of them illegal, and Europeans are very divided in their reaction to this phenomenon, as I note in the essay accompanying this chapter. Some see the immigrants as the necessary resource that will keep the economy running and pension checks flowing for aging Europeans. Others see the immigrants as a very real threat to the European way of life, coming as they mainly do from Africa and Asia.

The changing demographic fortunes of the global population is demonstrated by the fact that, as you can see in Table 2.2, in 1950 there were seven European countries in the world's top 20; by 2015 there were only two; and by 2050 it is projected that there will only be one.

IMPLOSION OR INVASION? THE CHOICES AHEAD FOR LOW-FERTILITY COUNTRIES

The world's population as a whole is in no danger of imploding anytime soon—quite the opposite. But the same cannot be said for much of Europe and parts of East Asia. Several countries in these areas are either already declining in population or are on the verge of doing so. The populations in Europe and East Asia all have birth rates that are below replacement level and have been that way for some time now, leading to a declining number of people at the younger ages. Further, it appears that the low fertility in these countries is not just a temporary phenomenon (Basten et al. 2013). Rather, it seems that the motivation to have large families has disappeared, at least for the time being, and has been replaced by a propensity to try to improve the family's standard of living by limiting the number of children. Russia has the added complication of having lost a few years of life expectancy over the past couple of decades, especially among men, further accelerating its population implosion. Most of the other Eastern European nations combine low fertility with the demographic complication that people are leaving to go elsewhere, primarily to Western Europe, but also to North America.

According to data from the United Nations Population Division, there are 17 countries estimated to have fewer people in 2015 than they had in 2010. Of these, 14 are in Eastern Europe, led by Russia and several of its former members of the Soviet Union, including Ukraine, Belarus, Georgia, Kazakhstan, the Republic of Moldova, Lithuania, Latvia, and Estonia. It is probably safe to say that the former Soviet Union has imploded. Germany is the only non–Eastern European country on the list, although as I note in this chapter its demographics are heavily influenced by having absorbed East Germany. The two other countries among these 17 are Japan and Cuba.

The most controversial issue surrounding current or impending population decline is that it is associated with an aging population. I will discuss the "mechanics" of population aging in Chapter 8, but the important point here is that the older population is increasing faster than the younger population and that has obvious economic implications. This is due partly to increasing life expectancy at the older ages (except in Russia), but mainly it is due to the very low birth rates in these countries. Furthermore, even if a nation's population is not yet declining numerically, the shift in numbers to an increasingly older population is still problematic to the extent that older people are generally "takers" rather than "contributors" to the economy.

One reaction to this situation is to suggest that this is a good thing for the planet as a whole, if not necessarily for countries involved. Residents of Europe (and Japan) are among the highest per-person consumers of the earth's resources, and if the populations eventually decline in size, their impact on the environment will be lower and the chance of global collapse thereby lessened (Diamond 2005). Within most of these countries, however, there is a concern about the economic impact of what many people are calling a "silver tsunami." Who will earn the money that is to be paid to retirees as pensions? Who will pay for the health care and social needs of the elderly? Who will keep the economy going so that the standard of living does not drop even as the expenses associated with population aging go up?

Several solutions to this dilemma have been proposed, and they relate to (1) raising the birth rate; (2) increasing labor force participation; and (3) possibly replacing the "missing" population with immigrants. In Chapter 6, I will discuss the fertility situation in some detail, but here we can note that countries with the lowest fertility rates are, in fact, those in which the least accommodation has been made to permit women to have a job and a family simultaneously. The availability of child care, programs for maternity leave and family leave, and societal pressure for men to help with child rearing and housework all increase the ability of women to participate in the labor force and still have children. Men have always had that ability, of course, but many countries, especially in Eastern and Southern Europe and East Asia, have opened up the labor market to women without making it easy to combine a woman's participation in the labor force with a family, and that has depressed birth rates below what they might otherwise be. Researchers have also noted that the effect of a low birth rate would be a little less troublesome if women simply had children at a younger age, even if they had the same number as they are currently having (Lutz et al. 2003). This adjustment would shorten the time between generations and would actually increase the growth rate by a slight amount.

The impact of an aging population on a nation's economy is exacerbated by an early age at retirement. For most of human history people simply worked until they were physically no longer able to do so. Retirement has been widely available for scarcely more than the past half century, but ever since that option was offered, people have been grabbing it. Guess what? Most people prefer retirement to work. Thus, we have witnessed the situation in which even as life expectancy has increased, people have been choosing to retire earlier. This wouldn't be a problem if all of these people had actually saved up enough money to live comfortably during a protracted retirement, but this is largely not the case. For the most part, people have been promised a retirement pension that is based on the transfer of money (through taxation) from people currently in the labor force to people who are retired (the so-called "pay-as-you-go" or PAYGO scheme). As long as the population was growing and the economy was improving, these promises were easy to keep (almost like a Ponzi scheme), but when these very same people who now want to collect a pension have not had enough children to supply the needs of the labor force, there is a problem.

One of the solutions involves raising the age at retirement. This has several benefits. It postpones the day when people will make a claim to a retirement pension, while at the same time keeping them in the labor force where they are economically productive and are continuing to pay taxes to help support those who are no longer working. This idea makes sense for at least two reasons: (1) with increasing education, people are entering the full-time labor force at older ages than prevailed when most old-age security schemes were put into place; and (2) life expectancy at the older ages is increasing, at the same time that physically exhausting jobs are on the decline, so most people are better able physically to stay in the labor force at older ages (Rau et al. 2013).

But the possibilities are even broader in scope. Vaupel and Kitowski (2008:255) note that "[w]ork hours must be spread more evenly over a longer life span. In this way, individuals will have the time necessary to bear and rear children and will be able to offer their expertise later in life. With such a policy, the elderly population would be occupied and supportive of society and youth would have the opportunity to conceive and care for children during those years in which they are physically able to do so." You can see that the beauty of this plan is that it addresses not only the problem of retiring at too young an age, but it also is designed to promote higher levels of childbearing among young couples by institutionalizing flexibility into the labor market.

Thus far, several European countries, including Germany and France, have raised the official retirement age, so there is some progress on this front, at least. The strongest motivation for governments to think about the broader changes to the work life course is largely to avoid the only other viable solution—import labor. This is, for example, the history of England, Germany, France, and several other European countries who needed labor to rebuild their economies after World War II. Between 1945 and the early 1970s, European nations allowed migration from former colonies, and they instituted guest worker programs, in which people contract to work for a few years and then go home again. The rub is that many workers choose not to go home. They stay and build families and become part of the fabric of their adopted society.

If workers came for a while, worked, and then left as they got older and were replaced by younger people, immigration wouldn't be too much of an issue. The Gulf States in the Middle East have managed to accomplish this largely by prohibiting workers from having families with them, and by forcing the deportation of workers who overstay their contract (Castles and Miller 2009). But Europeans have rarely been willing to take those extreme measures, so guest workers are likely to stay past the end of their contract to become undocumented immigrants. The reality, then, is that replacement migration in Europe

(continued)

IMPLOSION OR INVASION? THE CHOICES AHEAD FOR LOW-FERTILITY COUNTRIES (CONTINUED)

means the immigration of not just workers but also their families, and within a generation or two the children of immigrants can become a major force in the demographic makeup of the receiving countries. In the meantime, many of these immigrants are not in the labor force, and thus are neither working nor paying taxes, so they are not exactly "replacing" the missing young adults (Bijak et al. 2008).

France and the United Kingdom have both taken in significant numbers of permanent immigrants from former colonies and, as a result, neither one is projected to decline in population over the next several decades. But the fact that an estimated 10 percent of France's population is now Muslim has created a variety of political and social dilemmas for that country. Data from the European Social Survey indicate that immigration is viewed negatively throughout Europe, mostly because of the perception that immigrants undermine European culture. There is little bias against migrants from elsewhere in the European Union (EU). Rather, it is non-EU immigrants that create anxiety (Markaki and Longhi 2013).

Outside of Europe, we find that Japan, like other Asian countries, has an extremely restrictive immigration policy because of an explicit desire to preserve the country's ethnic homogeneity. Although Japan does tolerate a small number of immigrants, it is unlikely that they will soon allow an invasion to prevent their impending implosion, despite concerns that declining population may significantly harm the nation's economy.

Discussion Questions: (1) Do you think it is appropriate or even possible for Europeans to increase their birth rate in order to stave off depopulation? Why or why not? **(2)** Why do you think Europe is more worried about immigration than the United States, but at the same time less concerned than Japan?

Northern Africa and Western Asia

The areas of the world usually described as Northern Africa and Western Asia are very similar to the MENA (Middle East and North Africa) region which I discussed in some detail in the essay in Chapter 1. However, though Iran is part of MENA, it is technically in South Asia, not Western Asia, and there are a few countries north of MENA that were once part of the former Soviet Union, but are technically in Western Asia. Overall, Northern Africa and Western Asia have a combined population of 471 million as of 2015, which is projected to increase to 692 million by 2050. The region is characterized especially by the presence of Islam (with the obvious notable exception of Israel), and by being one of the more rapidly growing areas of the world, in which violence and conflict have all too often gone hand in hand with rapid growth and youth bulges.

Egypt is the most populous of the countries in Northern Africa and Western Asia, with 85 million people (fifteenth most populous in the world), followed closely by Turkey with 77 million (eighteenth most populous). Together they account for more than one-third of the region's total population. Egyptians are crowded into the narrow Nile Valley. With its rate of growth of 1.9 percent per year, Egypt's population would double in 36 years without a significant drop in the birth rate, and this rapid growth constantly hampers even the most ambitious strategies for economic growth and development. Indeed, this is almost certainly a key reason for the political turmoil in Egypt. As is true for nearly all countries in this region of the world, the explosive growth in numbers is due to the dramatic drop in mortality since the end of World War II. In 1937, the life expectancy at birth in Egypt was less than 40 years (Omran and Roudi 1993), whereas by now it has risen to 70. Even with such an improvement in mortality, however, death rates are just at the world average, while the number of children born to women (3.0) is well above the world average (2.5). Because of this high fertility, a very high proportion (31 percent) of the population is under age 15, and that is part of the recipe for the problems that beset Egypt.

It is the size and rate of increase in the youthful population that has been especially explosive throughout northern Africa and the Arab societies of western Asia, as I alluded to in the previous chapter. Somewhat presciently, prior to the Arab Spring, *The Economist* (2009:8) put it succinctly: "By far the biggest difficulty facing the Arabs—and the main item in the catalogue of socio-economic woes submitted as evidence of looming upheaval—is demography." The rapid drop in mortality after World War II, followed by a long delay in the start of fertility decline, produced a very large population of young people in need of jobs. They have spread throughout the region looking for work, and many have gone to Europe and North and South America. The economies within the region have not been able to keep up with the demand for jobs, and this has produced a generation of young people who, despite being better educated than their parents, face an uncertain future in an increasingly crowded world. The demographic situation has fueled discontent and has almost certainly contributed to the rise of radical Islam and terrorism.

Turkey has fared better demographically than most of the Arab nations that were once in its orbit when Turkey was the political center of the Ottoman Empire. Its fertility has recently dropped to replacement level and life expectancy is five years above the global average. The percentage of the population under 15 is steadily declining, and the overall rate of growth is sufficiently slow that the United Nations projects that by 2050 it will no longer be on the top 20 list. Its demography is edging closer to a European pattern, consistent with its push to join the European Union. At the same time, the western part of the Turkey (closer to Europe) is demographically more European than the eastern part of the country, where fertility is much higher and female literacy much lower (Isik and Pinarcioglu 2006; Courbage and Todd 2011). It is notable that Turkey's southeastern neighbors —Iraq and Syria—have high birth rates, high growth rates, high fractions under age 15, and high levels of conflict and violence. Unfortunately, these countries seem more typical of the region than does Turkey.

Sub-Saharan Africa

According to most evidence, Sub-Saharan Africa is the place from which all human life originated (see, for example, Wilson and Cann 1992), and the 949 million people living there now comprise 13 percent of the world's total. Nigeria, with 183 million (seventh most populous in the world) accounts for nearly 1 in 5 of those 949 million, followed by Ethiopia (thirteenth most populous) and the Democratic Republic of the Congo—nineteenth most populous in the world. Note that there are two countries with Congo in the name: the most populous (the Democratic

Republic) has Kinshasa as its capital, and the other less populous Republic of Congo has Brazzaville as its capital.

All three of these countries (as well as their neighbors) have incredibly high levels of fertility, especially considering the fact that death rates are much lower than they used to be. In both Nigeria and the Congo the average woman is currently having 6 children, while in Ethiopia the average is 4.5 children. Not surprisingly, these high birth rates, in combination with declining infant and child mortality, produce young populations. All three of these countries have very close to 45 percent of the population under age 15. This means rapid population growth, and Nigeria is projected to vault over the United States as the third most populous country by the middle of this century. Ethiopia is projected to move into the top 10, and the Congo will be close behind. Their neighbors, Tanzania, Uganda, and Kenya, are also projected to move into the top 20 by 2050.

South and Southeast Asia

South and Southeast Asia as a region is home to 2.4 billion people, one-third of the world's total. The Indian subcontinent dominates this area demographically—India (the world's second most populous nation), Pakistan (sixth), and Bangladesh (eighth) encompass two-thirds of the region's population. But Indonesia, the world's fourth most populous nation (and the one with the largest Muslim population in the world), is also part of Southeast Asia, as are three other countries on the top 20 list—the Philippines (twelfth), Vietnam (fourteenth), and Thailand (twentieth). And we cannot forget that Iran, the world's seventeenth most populous nation, is technically in South Asia, even though also in MENA, as I noted previously.

India, Pakistan, and Bangladesh Second to China in population size, at least for the moment, is India, with the current population estimated to be 1.3 billion, but projected to be 1.6 billion (more populous than China) by the middle of this century (see Table 2.2). Mortality is somewhat higher in India than in China, and the birth rate is quite a bit higher. Indian females have a life expectancy at birth of 68 years—five years below the world average, but a substantial improvement over the 27 years that prevailed back in the 1920s (Adlahka and Banister 1995). The infant mortality rate of 44 per 1,000 is higher than the world average, but it is also far lower than it was just a few decades ago. Women are bearing children at a rate of 2.4 each, and most children in India now survive to adulthood. With an annual growth rate of 1.5 percent, the Indian population is increasing by 19 million people each year. Thus, nearly one in four people being added to the world's population annually is from India. The population of the Indian subcontinent is already more populous than mainland China, and that does not take into account the millions of people of Indian and Pakistani origin who are living elsewhere in the world.

India's population is culturally diverse, and this is reflected in rather dramatic geographic differences in fertility and rates of population growth within the country. In the southern states of Kerala and Tamil Nadu, fertility had dropped below the replacement level by the mid-1990s and has stayed there since. However, in the four most populous states in the north (Bihar, Madhya Pradesh, Rajasthan, and Uttar Pradesh), where 40 percent of the Indian population lives, the average woman was bearing more than three children, according to fertility survey data (Measure DHS 2014).

At the end of World War II, when India was granted its independence from British rule, the country was divided into predominantly Hindu India and predominantly Muslim Pakistan, with the latter having territory divided between West Pakistan and East Pakistan. In 1971, a civil war erupted between the two disconnected Pakistans, and, with the help of India, East Pakistan won the war and became Bangladesh. Although Pakistan and Bangladesh are both Muslim, Bangladesh has a demographic profile that now looks more like India than Pakistan. The average woman in Bangladesh now gives birth to 2.3 children (slightly fewer than in India), whereas fertility in Pakistan has remained much higher (currently 3.8 children per woman). The overall rate of population growth in Bangladesh is 1.5 percent per year (exactly the same as India's), but it is 2.3 percent per year in Pakistan. Still, both Pakistan and Bangladesh have grown so much since independence in 1947 that, were they still one country, they would be the third most populous nation in the world.

Indonesia and the Philippines Indonesia is a string of nearly 18,000 islands in Southeast Asia, with an estimated 256 million people spread out among nearly 1,000 of those islands. A former Dutch colony, it has experienced a substantial decline in fertility in recent years, but Indonesian women nonetheless are bearing an above-average level of 2.6 children each. Given the increasing life expectancy, now just a year below the world average, the population is growing at 1.5 percent per year—similar to India and Bangladesh. For several decades, Indonesia has dealt with population growth through a program of transmigration, in which people have been sent from the more populous to the less populous islands. These largely forested outer islands have suffered environmentally from the human encroachment, without necessarily dealing successfully with Indonesia's basic dilemma, which is how to raise its burgeoning young adult population out of poverty. This dilemma has contributed to increased political instability, as well as a rise in the level of Islamic fundamentalism and terrorism.

The Philippines is a set of more than 7,000 islands to the north of Indonesia. About 2,000 of those islands are inhabited. It has even higher fertility than Indonesia (an average of 3.0 children per woman), but also experiences more outmigration than does Indonesia. This may relieve some of the pressure felt in the Philippines by the fact of having a large youth population, but the country is still struggling under the weight of its demographic growth. Although the country is predominantly Catholic, concentrated especially in the northern Luzon islands, there have long been clashes with Muslims in the southern group of islands comprising Mindanao. These ethno-religious differences are reflected in the demographic trends, with lower fertility and child mortality in the Luzon region than in the Mindanao region.

Vietnam and Thailand Back on the Asian mainland, we find that Vietnam and Thailand have both boomed demographically since the days when the United States was involved in this region's conflict. Vietnam, in particular, has nearly doubled in population since the Vietnam War ended in 1975. That was largely a result of a swift drop in mortality unaccompanied immediately by a decline in fertility, thus leading to a huge youth bulge. Recognizing the threat to development, Vietnam introduced a national family planning policy in 1988 encouraging (although not forcing) couples to have only one or two children (Goodkind 1995). That policy, in concert with Chinese-style free-market economic reforms, led to a swift drop in the number of children per woman—from an average of 6 each in 1975 to replacement level in 2000, where it has stayed. The result has been the same kind of "demographic dividend" that China has experienced, as I note below.

Fertility and mortality both dropped sooner in Thailand than in Vietnam. As early as the 1990s, fertility levels were below replacement in the capital of Bangkok, and the rest of the country quickly followed suit. However, as in Vietnam, the swift decline in fertility meant that there was a huge youth bulge that swelled the population, even though at an individual level women were not bearing many children. Now, however, the country is aging and is expected to have fewer people in 2050 than now, thereby dropping off the top 20 list by mid-century.

Iran With 79 million people, Iran is the most populous Shia-majority Muslim country in the world (followed by Iraq, Azerbaijan, and Bahrain—the only other Shia-majority countries). Like Vietnam and Thailand, it has experienced a very rapid fertility decline—from an average of 6 children per woman as recently as 1985 to below replacement level today (Lutz et al. 2010; Courbage and Todd 2011). This has gone hand in hand with a rise in female literacy and other forms of modernization taking place in the country, despite the generally conservative attitude of the government. Here again we see a population that now has low fertility and high life expectancy but which is still experiencing fairly rapid population growth because of the momentum built into the large youth bulge created by the high fertility of the recent past. I will return to the theme of population momentum in Chapter 8 in the discussion of the age transition.

East Asia

East Asia has 1.6 billion people, with the region dominated demographically by China, the most populous country in the world with 1.4 billion people, and Japan, the tenth most populous even though its 127 million is less than ten percent of China's size. Overall, East Asia includes more than 20 percent of the world's total population, but its share is diminishing as China continues to brake its population growth and as Japan teeters on the edge of depopulation. Indeed, these two countries, along with South Korea and Taiwan (the other major countries in the region) are projected to have fewer people in 2050 than they do now.

China With one-fifth of all human beings, The People's Republic of China dominates the map of the world drawn to scale according to population size (see Figure 2.1). If we add in the Chinese in Taiwan (which the government of mainland China still claims as its own), Singapore, and the overseas Chinese elsewhere in the world, closer to one out of every four people is of Chinese origin. Nonetheless, China's share of the world's total population actually peaked in the middle of the nineteenth century. In 1850, more than one in three people were living in China, and that fraction has steadily declined over time, even as China's population continued to grow in absolute numbers, fueled by high birth rates that tended to compensate for the high death rates.

After the communist overthrow of China in 1949, the government at first tried to ignore the country's demographic bulk, partly for Marxist ideological reasons that I will discuss in the next chapter. However, after the death of Mao Zedong in 1976, China began to take stock of the magnitude of its problem. In 1982 it conducted its first national census since 1964 (which had been taken shortly after the terrible famine that I mentioned earlier). Fertility had begun to decline in earnest by then, as I will discuss in more detail in Chapter 6, but nonetheless the census counted more than 1 billion people. The general government attitude was summed up in the mid-1990s as follows:

> Despite the outstanding achievements made in population and development, China still confronts a series of basic problems including a large population base, insufficient cultivated land, under-development, inadequate resources on a per capita basis and an uneven social and economic development among regions. . . . Too many people has [sic] impeded seriously the speed of social and economic development of the country and the rise of the standard of living of the people. Many difficulties encountered in the course of social and economic development are directly attributable to population problems. (Peng 1996:7)

Fertility decline actually began in China's cities in the 1960s and spread rapidly throughout the rest of the country in the 1970s, when the government introduced the family planning program known as *wan xi shao*, meaning "later" (marriage), "longer" (birth interval), "fewer" (children) (Goldstein and Feng 1996). In 1979, this was transformed into the now famous (if not infamous) one-child policy, but fertility was already on its way down by that time (Riley 2004). Indeed, it dropped to replacement level in the 1990s and has stayed there since.

Although China's birthrate has now dropped to 1.5 children per woman, that does not yet mean that the population has stopped growing—population momentum again rears its head. Despite its low birth rate, the number of births each year in China is nearly twice the number of deaths just because China is paying for its previous high birth rate. There are so many young women of reproductive age (women born 25 to 45 years ago when birth rates were still above replacement level) that their babies still outnumber the people who are dying each year. As a result, the rate of natural increase in China is essentially the same as in the United States, despite the lower fertility rate.

China has famously used its "demographic dividend" (a bulge of adults unencumbered by a lot of children due to the rapid decline in fertility) to create jobs and grow its economy. But this is just a transition period for China between its formerly very young population and its quickly aging population. Thus, population growth remains a serious concern in China, but the concern is now turning from the young population to the rapidly increasing number and proportion of older Chinese—the inevitable consequence of a rapid decline in fertility in a nation where mortality is also low. China may be unique in the world in "getting old" before it has gotten rich. Despite a loosening of the one-child policy by the government in 2013, no one anticipates a huge increase in China's birth rate, and it is projected to decline in population between now and the middle of the century, yielding its long-held position of most populous to India.

Japan Population size probably peaked in Japan in 2010 and is now slowly on the way down. The decline is actually slower than it might be due to the fact Japan has the lowest level of mortality in the world, with a female life expectancy at birth of 86 years. Japan's health (accompanied by its wealth) translates demographically into very high probabilities of survival to old age—indeed, more than half of all Japanese born this year will likely still be alive at age 80. This very low mortality rate is accompanied by very low fertility. Japanese women are bearing an average of 1.4 children each, leading some pundits to suggest that Japan has its own "one-child policy." Japan's low mortality and low fertility have produced a population in which only 13 percent are under age 15, whereas 26 percent are 65 or older. The United Nations forecasts that by 2050 the percent under 15 will have dropped still slightly to 12 percent, while the percent 65 and older will have jumped to 37. This will be associated with a projected population decline from the current 127 million down to 108 million, although it will still be in the top 20 in 2050, as you can see in Table 2.2.

Oceania

None of the countries in Oceania is populous enough to be on the top 20 list. It is home to a wide range of indigenous populations, including Melanesian and Polynesian, but European influence has been very strong, and the region is generally thought of as being "overseas European." Its population of 39 million is just slightly more than Canada's, and is less than 1 percent of the world's total. Australia accounts for two-thirds of the region's population, followed by Papua New Guinea and New Zealand.

In a pattern repeated elsewhere in the world, the lowest birth rates and lowest death rates (and thus the lowest rates of population growth) are found in countries whose populations are largely European-origin (Australia and New Zealand, in this case); whereas the countries with a higher fraction of the population of indigenous origin have higher birth rates, higher mortality, and substantially higher rates of population growth (exemplified in Oceania by Papua New Guinea). Much of Australia's population growth is fueled by immigration, especially from Asia, and this is leading to a situation not unlike that in the United States, where an increasingly older European-origin population is supported by a younger generation of children of immigrants.

This whirlwind global tour highlights the tremendous demographic contrasts that exist in the modern world. In the less developed nations, the population continues to grow quickly, especially in absolute terms, not just in terms of rates of growth. In sub-Saharan Africa this is happening even in the face of the HIV/AIDS pandemic. Yet, in the more developed countries population growth has slowed, stopped, or in some places even started to decline. As we look around the world, we see that the more rapidly growing countries tend to have high proportions of people who are young, poor, prone to disease, and susceptible to political instability. The countries that are growing slowly or not at all tend to have populations that are older, richer, and healthier, and these are the nations that are politically more stable.

There is almost certainly something to the idea that "demography is destiny"—a country cannot readily escape the demographic changes put into motion by the universally sought-after decline in mortality. Each country has to learn how to read its own demographic situation and cope as well as it can with the inevitable changes that will take place as it evolves through all phases of the demographic transition.

Summary and Conclusion

High death rates kept the number of people in the world from growing rapidly until approximately the time of the Industrial Revolution. Then improved living conditions, public health measures, and, more recently, medical advances dramatically accelerated the pace of growth. As populations have grown, the pressure or desire to migrate has also increased. The vast European expansion into less developed areas of the world, which began in the fifteenth and sixteenth centuries but accelerated in the nineteenth century, is a notable illustration of massive migration and population redistribution. Today migration patterns have shifted, and people are mainly moving from less developed to more developed nations. Closely associated with migration and population density is the urban revolution—that is, the movement from rural to urban areas.

The current world situation finds China and India to be the most populous countries, followed by the United States, Indonesia, and Brazil. Everywhere population is growing we find that death rates have declined more rapidly than have birth rates, but there is considerable global and regional variability in both the birth and death rates and thus in the rate of population growth. Dealing with the pressure of an expanding young population is the task of developing countries; whereas more developed countries, along with China, have aging populations and are coping with the fact that the demand for labor in their economies may have to be met by immigrants from more rapidly growing countries.

Demographic dynamics represent the leading edge of social change in the modern world. It is a world of more than 7 billion people, heading to more than 9 billion by the middle of this century and probably even more beyond that.

In order to cope with these demographic underpinnings of our lives, we need to have a demographic perspective that allows us to sort out the causes and consequences of population change. We turn to that in the next chapter.

Main Points

1. During the first 90 percent of human existence, the population of the world had grown only to the size of today's New York City.

2. Between 1750 and 1950, the world's population mushroomed from 800 million to 2.5 billion, and since 1950 it has expanded to more than 7 billion.
3. Despite the fact that humans have been around for hundreds of thousands of years, more than one in ten people ever born is currently alive.
4. Early population growth was slow not because birth rates were low but because death rates were high; on the other hand, continuing population increases are due to dramatic declines in mortality without a matching decline in fertility.
5. World population growth has been accompanied by migration from rapidly growing areas into less rapidly growing regions. Initially, that meant an outward expansion of the European population, but more recently it has meant migration from less developed to more developed nations.
6. Migration has also involved the shift of people from rural to urban areas, and urban regions on average are currently growing more rapidly than ever before in history.
7. Although migration is crucial to the demographic history of the United States and Canada, both countries have grown largely as a result of natural increase—the excess of births over deaths—after the migrants arrived.
8. At the time of the American Revolution, fertility levels in North America were among the highest in the world. Now they are low, although not as low as in Europe.
9. The world's 10 most populous countries are the People's Republic of China, India, the United States, Indonesia, Brazil, Pakistan, Nigeria, Bangladesh, Russia, and Japan. Together they account for 59 percent of the world's population.
10. Almost all of the population growth in the world today is occurring in the less developed nations, leading to an increase in the global demographic contrasts among countries.

Questions for Review

1. Describe what you think might be the typical day in the life of a person living in a world where death rates and birth rates are both very high. How might those demographic imperatives influence everyday life? How would "culture" be different from today as a result?
2. The media in the United States and Europe regularly have stories about the impact of low fertility slowing down population growth in these countries. If you were asked to be on a TV talk show commenting on such a story, how would you respond?
3. Migration of people into other countries is a major part of the demography of the modern world. How do you think the world of 2050 will look demographically as a consequence of the trends currently in place?

4. Even without migration, the world will look very different in 2050 than it did in 1950. Analyze Table 2.2 in terms of the idea that "the past is a foreign country."
5. How would you explain the regional patterns that are very observable with respect to global demography? Are European countries more like each other than they are like Asian countries? Is Africa unique demographically? Are national boundaries therefore meaningless when it comes to population trends?

Websites of Interest

Remember that websites are not as permanent as books and journals, so I cannot guarantee that each of the following websites still exists at the moment you are reading this. You may have to Google the name of the organization to find the current web address.

1. **http://www.gapminder.org/videos/dont-panic-the-facts-about-population/**
   Hans Rosling is a Swedish academic—Professor of International Health at Karolinska Institute in Sweden and co-founder and chairman of the Gapminder Foundation (gapminder.org). In this program prepared for BBC in 2013, he talks about the current world population situation. Check out his other population-related talks because he does a nice job of explaining things visually.
2. **http://censusindia.gov.in**
   You don't have to take anybody else's word for what's happening demographically in India. This Indian census website is in English and has lots of data for the country and its regions.
3. **http://sedac.ciesin.columbia.edu/data/collection/gpw-v3**
   The Gridded Population of the World is a database created from censuses, surveys, satellite imagery, and other sources, producing a very realistic picture of population density and other characteristics at the global level. Regional maps and data are also available at this website.
4. **http://www.worldpop.org.uk**
   The WorldPop project was initiated in October 2013 to combine the AfriPop, AsiaPop, and AmeriPop population mapping projects. It aims to provide an open access archive of spatial demographic datasets for Central and South America, Africa and Asia to support development and health applications. The methods used are designed with full open access and operational application in mind, using transparent, fully documented and shareable methods to produce easily updatable maps with accompanying metadata.
5. **http://weekspopulation.blogspot.com/search/label/Global%20Population%20Trends**
   Keep track of the latest news related to this chapter by visiting my WeeksPopulation website.

CHAPTER 3
Demographic Perspectives

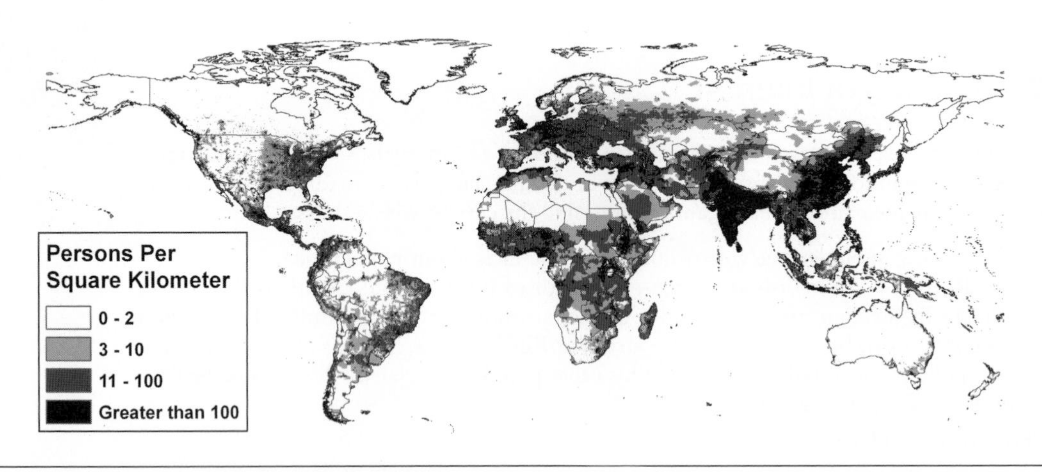

Figure 3.1 Population Density of the World

Source: Adapted by the author from Center for International Earth Science Information Network - CIESIN - Columbia University, and Centro Internacional de Agricultura Tropical - CIAT. 2005. Gridded Population of the World, Version 3 (GPWv3): Population Density Grid, Future Estimates. Palisades, NY: NASA Socioeconomic Data and Applications Center (SEDAC). http://sedac.ciesin.columbia.edu/data/set/gpw-v3-population-density-future-estimates. Accessed 2014. Data are based on United Nations Population Division projections for the year 2015.

PREMODERN POPULATION DOCTRINES

THE PRELUDE TO MALTHUS

THE MALTHUSIAN PERSPECTIVE
Causes of Population Growth
Consequences of Population Growth
Avoiding the Consequences
Critique of Malthus
Neo-Malthusians

THE MARXIAN PERSPECTIVE
Causes of Population Growth
Consequences of Population Growth
Critique of Marx

THE PRELUDE TO THE DEMOGRAPHIC TRANSITION THEORY
Mill
Dumont
Durkheim

THE THEORY OF THE DEMOGRAPHIC TRANSITION
Critique of the Demographic Transition Theory
Reformulation of the Demographic Transition Theory
The Theory of Demographic Change and Response
Cohort Size Effects
Is There Something Beyond the Demographic Transition?

THE DEMOGRAPHIC TRANSITION IS REALLY A SET OF TRANSITIONS
The Health and Mortality Transition
The Fertility Transition
The Age Transition
The Migration Transition
The Urban Transition
The Family and Household Transition
Impact on Society

ESSAY: Who Are the Neo-Malthusians?

To get a handle on population problems and issues, you have to put the facts of population together with the "whys" and "wherefores." In other words, you need a **demographic perspective**—a way of relating basic information to theories about how the world operates demographically. A demographic perspective will guide you through the sometimes tangled relationships between population factors (such as size and growth, geographic distribution, age structure, and other sociodemographic characteristics) and the rest of what is going on in society. As you develop your demographic perspective, you will acquire a new awareness about your own community, as well as about national and world political, economic, and social issues. You will be able to understand the influences that demographic changes have had (or might have had), and you will consider the demographic consequences of events, which historically have been huge and transformative.

In this chapter, I discuss several theories of how population processes are entwined with general social processes. There are actually two levels of population theory. At the core of demographic analysis is the technical side of the field—the mathematical and biomedical theories that predict the kinds of changes taking place in the biological components of demography: fertility, mortality, and the distribution of a population by age and sex. Demography, for example, has played a central role in the development of the fields of probability, statistics, and sampling (Kreager 1993). This hard core is crucial to our understanding of human populations, but there is a "softer" (although no less important) outer wrapping of theory that relates demographic processes to the real events of the social world (Schofield and Coleman 1986). The linkage of the core with its outer wrapping is what produces a demographic perspective.

Two questions have to be answered before you will be able to develop your own perspective: (1) What are the *causes* of population change; and (2) What are the *consequences* of population change? In this chapter, I discuss several perspectives that provide broad answers to these questions and that also introduce the major lines of demographic theory. The purpose of this review is to give you a start in developing your own demographic perspective by taking advantage of what others have learned and passed on to us.

I begin this chapter with a brief review of premodern thinking on the subject of population. Most of these ideas are what we call **doctrine**, as opposed to **theory**. Early thinkers were certain they had the answers and certain that their proclamations represented the truth about population growth and its implications for society. By contrast, the essence of modern scientific thought is to assume that you do not have the answer and to acknowledge that you are willing to consider evidence regardless of the conclusion to which it points. In the process of sorting out the evidence, we develop tentative explanations (hypotheses and then theories) that help guide our thinking and our search for understanding. In demography, as in all of the sciences, theories replace doctrine when new, systematically collected information (censuses and other sources discussed in the next chapter) becomes available, allowing people to question old ideas and formulate new ones. Table 3.1 summarizes the doctrines and theories discussed in the chapter.

Table 3.1 Demographic Perspectives over Time

	Date	Demographic Perspective
Examples of Premodern Doctrines	~1300 B.C.	Book of Genesis—"Be fruitful and multiply."
	~500 B.C.	Confucius—Population growth is good, but governments should maintain a balance between population and resources.
	360 B.C.	Plato—Population quality more important than quantity; emphasis on population stability.
	340 B.C.	Aristotle—Population size should be limited and the use of abortion might be appropriate.
	~50 B.C.	Cicero—Population growth necessary to maintain Roman influence
	A.D. 400	St. Augustine—Abstinence is the preferred way to deal with human sexuality; the second best is to marry and procreate.
	A.D. 1280	St. Thomas Aquinas—Celibacy is *not* better than marriage and procreation.
	A.D. 1380	Ibn Khaldun—Population growth is inherently good because it increases occupational specialization and raises incomes.
	1500–1800	Mercantilism—Increasing national wealth depends on a growing population that can stimulate export trade
	1700–1800	Physiocrats—Wealth of a nation is in land, not people; therefore population size depends on the wealth of the land, which is stimulated by free trade (laissez-faire).
Modern Theories	1798	Malthus—Population grows exponentially while food supply grows arithmetically, with misery (poverty) being the result in the absence of moral restraint
	~1800	Neo-Malthusian—Accepting the basic Malthusian premise that population growth tends to outstrip resources, but unlike Malthus believing that birth control measures are appropriate checks to population growth.
	~1844	Marxian—Each society at each point in history has its own law of population that determines the consequences of population growth; poverty is not the natural result of population growth.

Table 3.1 (continued)

Date	Demographic Perspective
~1873 to 1929	Prelude to the demographic transition, including Mill, Dumont, Durkheim, and Thompson
1945	Demographic transition in its original formulation—The process whereby a country moves from high birth and death rates to low birth and death rates with an interstitial spurt in population growth. Explanations based originally on modernization theory.
1962	Earliest studies suggesting the need to reformulate the demographic transition theory.
1963	Theory of demographic change and response—Demographic response made by individuals to population pressures is determined by the means available to them to respond; causes and consequences of population change are intertwined.
1968	Easterlin relative cohort size hypothesis—Successively larger young cohorts put pressure on young men's relative wages, forcing them to make a tradeoff between family size and overall well-being.
1971–present	Decomposition of the demographic transition into its separate transitions—Health and mortality, fertility, age, migration, urbanization, and family and household.

Premodern Population Doctrines

Until about 2,500 years ago, human societies probably shared a common concern about population: They valued reproduction as a means of replacing people lost through universally high mortality. Ancient Judaism, for example, provided the prescription to "be fruitful and multiply" (Genesis 1:28). Indeed, reproductive power was often deified, as in ancient Greece, where it was the job of a variety of goddesses to help mortals successfully bring children into the world and raise those children to adulthood. In two of the more developed areas of the world 2,500 years ago, however, awareness of the potential for populations to grow beyond their resources prompted comment by well-known philosophers. In the fifth century B.C., the writings of the school of Confucius in China discussed the relationship between population and resources (Sauvy 1969), and it was suggested that the government should move people from overpopulated to underpopulated areas (an idea embraced in the twentieth century by the Indonesian government). Nonetheless, the idea of promoting population growth was clear in the doctrine of Confucius (Keyfitz 1973).

Plato, writing in *The Laws* in 360 B.C., emphasized the importance of population stability rather than growth. Specifically, Plato proposed keeping the ideal community of free citizens (as differentiated from indentured laborers or slaves who had few civil rights) at a constant 5,040. Charbit (2002:216) suggests that "what inspired Plato in his choice of 5,040 is above all the fact that it is divisible by twelve, a number with a decisive sacred dimension," a legacy carried on in the 12 months of the year, among dozens (pun intended) of other aspects of modern life. The number of people desired by Plato was still moderately small, because Plato felt that too many people led to anonymity, which would undermine democracy, whereas too few people would prevent an adequate division of labor and would not allow a community to be properly defended. Population size would be controlled by late marriage, infanticide, and migration (in or out as the situation demanded) (Plato 360 B.C. [1960]). Plato was an early proponent of the doctrine that quality in humans is more important than quantity.

In the Roman Empire, the reigns of Julius and Augustus Caesar were marked by clearly pronatalist doctrines—a necessity, given the very high mortality that characterized the Roman era (Frier 1999). In approximately 50 B.C., Cicero noted that population growth was seen by the leaders of Rome as a necessary means of replacing war casualties and of ensuring enough people to help colonize new lands. Several scholars have speculated, however, that by the second century A.D., as the old, pagan Roman empire was waning in power, the birth rate in Rome may have been declining (Stangeland 1904; Veyne 1987). In a thoroughly modern sentiment, Pliny ("the younger") complained that ". . . in our time most people hold that an only son is already a heavy burden and that it is advantageous not to be overburdened with posterity" (quoted in Veyne 1987:13).

The Middle Ages in Europe, which followed the decline of Rome and its transformation from a pagan to a Christian society, were characterized by a combination of both **pronatalist** and **antinatalist** Christian doctrines. Christianity condemned polygamy, divorce, abortion, and infanticide—practices that had kept earlier Roman growth rates lower than they otherwise might have been. The early and highly influential Christian leader, mystic, and writer Augustine (A.D. 354–430) interpreted the message of Paul in the New Testament to mean that virgins were the highest form of human existence. Human sexuality was, in Augustine's view, a supernaturally good thing but also an important cause of sin (because most people are unable or unwilling to control their desires) (O'Donnell 2006). He believed that abstinence was the best way to deal with sexuality (an antinatalist view), but the second-best state was marriage, which existed for the purpose of procreation (a pronatalist view). This duality shows up most clearly in the Roman Catholic Church where priests take a vow of abstinence while urging their parishioners to have as many children as possible.

The time between the end of the Roman Empire (fifth century A.D.) and the Renaissance (fifteenth century A.D.) was an economically stagnant, fatalistic period of European history. While Europe muddled through the Middle Ages, Islam (which had emerged in the seventh century A.D.) was expanding throughout the Mediterranean. Muslims took control of southern Italy and the Iberian peninsula and, under the Ottoman Empire, controlled the Balkans and the rest of southeastern Europe. Europe's reaction to this situation was the Crusades, a series of wars launched by Christians to wrestle control away from Muslims. These expeditions were largely unsuccessful from a military perspective, but they did put Europeans into contact with the Muslim world, which ultimately led to the Renaissance—the rebirth of Europe:

> The Islamic contribution to Europe is enormous, both of its own creations and of its borrowings—reworked and adapted—from the ancient civilizations of the eastern Mediterranean and from the remoter cultures of Asia. Greek science and philosophy, preserved and improved by the Muslims but forgotten in Europe; Indian numbers and Chinese paper; oranges and lemons, cotton and sugar, and a whole series of other plants along with the methods of cultivating them—all these are but a few of the many things that medieval Europe learned or acquired from the vastly more advanced and more sophisticated civilization of the Mediterranean Islamic world. (Lewis 1995:274)

By the fourteenth century, one of the great Arab historians and philosophers, Ibn Khaldun, was in Tunis writing about the benefits of a growing population. In particular, he argued that population growth creates the need for specialization of occupations, which in turn leads to higher incomes, concentrated especially in cities: "Thus, the inhabitants of a more populous city are more prosperous than their counterparts in a less populous one. . . . The fundamental cause of this is the difference in the nature of the occupations carried on in the different places. For each town is a market for different kinds of labour, and each market absorbs a total expenditure proportionate to its size" (quoted in Issawi 1987:268). Ibn Khaldun was not a utopian. His philosophy was that societies evolved and were transformed as part of natural and normal processes. One of these processes was that "procreation is stimulated by high hopes and resulting heightening of animal energies" (quoted in Issawi 1987:268).

To be sure, the cultural reawakening of Europe took place in the context of a growing population, as I noted in the previous chapter. Not surprisingly, then, new murmurings were heard about the place of population growth in the human scheme of things. The Renaissance began with the Venetians, who had established trade with Muslims and others as the eastern Mediterranean ceased to be a Crusade war zone in the thirteenth century. In that century, an influential Dominican monk, Thomas Aquinas, argued that marriage and family building were not inferior to celibacy, thus implicitly promoting the idea that population growth is an inherently good thing.

By the end of the fourteenth century, the plague had receded from Europe; by the sixteenth century, Muslims (and Jews) had been expelled from southern Spain, and Europeans had begun their discovery and exploitation of Africa, the Americas, and south Asia. Cities began to grow noticeably, and Giovanni Botero, a sixteenth-century Italian statesman, wrote that "the powers of generation are the same now as one thousand years ago, and, if they had no impediment, the propagation of man would grow without limit and the growth of cities would never stop" (quoted in Hutchinson 1967:111). The seventeenth and eighteenth centuries witnessed an historically unprecedented trade (the so-called **Columbian Exchange**) of food, manufactured goods, people, and disease between the Americas and most of the rest of the world (Crosby 1972), undertaken largely by European merchants, who had the best ships and the deadliest weapons in the world (Cipolla 1965; Diamond 1997).

This rise in trade, prompted at least in part by population growth, generated the doctrine of **Mercantilism** among the new nation-states of Europe. Mercantilism maintained that a nation's wealth was determined by the amount of precious metals it had in its possession, which were acquired by exporting more goods than were imported, with the difference (the profit) being stored in precious metals. The catch here was that a nation had to have things to produce to sell to others, and the idea was that the more workers you had, the more you could produce. Furthermore, if you could populate the new colonies, you would have a ready-made market for your products. Thus population growth was seen as essential to an increase in national revenue, and Mercantilist writers sought to encourage it by a number of means, including penalties for non-marriage, encouragements to get married, lessening penalties for illegitimate births, limiting out-migration (except to their own colonies), and promoting immigration of productive laborers. It is important to keep in mind that these doctrines were concerned with the wealth and welfare of a specific country, not all of human society. "The underlying doctrine was, either tacitly or explicitly, that the nation which became the strongest in material goods and in men would survive; the nations which lost in the economic struggle would have their populations reduced by want, or they would be forced to resort to war, in which their chances of success would be small" (Stangeland 1904:183).

Mercantilist doctrines were supported by the emerging demographic analyses of people like John Graunt, William Petty, and Edmund Halley (all English) in the seventeenth century, and Johann Peter Süssmilch, an eighteenth-century chaplain in the army of Frederick the Great of Prussia (now Germany). In 1662, John Graunt, a Londoner who is sometimes called the father of demography, analyzed the series of Bills of Mortality in the first known statistical analysis of demographic data (Sutherland 1963). Although he was a haberdasher by trade, Graunt used his spare moments to conduct studies that were truly remarkable for his time. He discovered that for every 100 people born in London, only 16 were still alive at age 36 and only 3 at age 66 (Graunt 1662 [1939]; Dublin et al. 1949)—suggesting very high levels of mortality. With these data he uncovered the high incidence of infant mortality in London and found, somewhat to the amazement of people at the time, that there were regular patterns of death in different parts of London. Graunt "opened the way both for the later discovery of uniformities in many social or volitional phenomena like marriage, suicide, and crime, and for a study of these uniformities, their nature and their limits; thus he, more than any other man, was the founder of statistics" (Willcox 1936:xiii). Indeed, Harrison and Carroll (2005) note that Graunt's studies are thought by many people to mark the beginning of social science as we know it today, not just statistics or demography.

One of Graunt's close friends (and probably the person who coaxed him into this work) was William Petty, a member of the Royal Society in London (Kreager 1988) and arguably the man who invented the field of economics (*The Economist* 2013). Petty circulated Graunt's work to the Society (which would not have otherwise paid much attention to a "tradesman"), and this brought it to the attention of the emerging scientific world of seventeenth-century Europe. Several years later, in 1693, Edmund Halley (of Halley's comet fame) became the first scientist to elaborate on the probabilities of death. Although Halley, like Graunt, was a Londoner, he came across a list of births and deaths kept for the city of Breslau in Silesia (now Poland). From these data, Halley used the life-table technique (discussed in Chapter 5) to determine that the expectation of life in Breslau between 1687 and 1691 was 33.5 years (Dublin et al. 1949).

Then, in the eighteenth century, Süssmilch built on the work of Graunt and others and added his own analyses to the observation of the regular patterns of marriage, birth, and death in Prussia and believed that he saw in these the divine hand of God ruling human society (Hecht 1987), in much the same way that people are fascinated by patterns such as the Fibonacci sequence. His view, widely disseminated throughout Europe, was that a larger population was always better than a smaller one, and, in direct contradistinction to Plato, he valued quantity over quality. He believed that indefinite improvements in agriculture and industry would postpone overpopulation so far into the future that it wouldn't matter.

The issue of population growth was more than idle speculation, because we know with a fair amount of certainty that the population of England, for example, doubled during the eighteenth century (Petersen 1979), and as I discussed in the previous chapter, Europe as a whole was increasing in population. The rising interest in population encouraged the publication of two important essays on population size, one by David Hume (1752 [1963]) and the other by Robert Wallace (1761 [1969]), which were then to influence Malthus, whom I discuss later.

These essays sparked considerable debate and controversy, because there were big issues at stake: "Was a large and rapidly growing population a sure sign of a society's good health? On balance, were the growth of industry and cities, the movement of larger numbers from one social class to another—in short, all of what we now term 'modernization'—a boon to the people or the contrary? And in society's efforts to resolve such dilemmas, could it depend on the sum of individuals' self-interest or was considerable state control called for?" (Petersen 1979:139). These are questions we are still dealing with 250 years later.

The population had, in fact, increased during the Mercantilist era, although probably not as a result of any of the policies put forth by its adherents. However, it was less obvious that the population was better off. Rather, the Mercantilist period had become associated with a rising level of poverty (Keyfitz 1972). Mercantilism relied on a state-sponsored system of promoting foreign trade, while inhibiting imports and thus competition. This generated wealth for a small elite but not for most people.

One of the more famous reactions against Mercantilism was that mounted in the middle of the eighteenth century by François Quesnay, a physician in the court of King Louis XV of France (and an economist when not "on duty"). Whereas Mercantilists argued that wealth depends on the number of people, Quesnay turned that around and argued that the number of people depends on the means of subsistence (a general term for level of living). The essence of this view, called **physiocratic** thought, was that land, not people, is the real source of wealth of a nation. In other words, population went from being an independent variable, causing change in society, to a dependent variable, being altered by societal change. As you will see throughout this book, both perspectives have their merits.

Physiocrats also believed that free trade (rather than the import restrictions demanded by Mercantilists) was essential to economic prosperity. This concept of "laissez-faire" (let people do as they choose) was picked up by Adam Smith, a Scotsman and one of the first modern economic theorists. Central to Smith's view of the world was the idea that, if left to their own devices, people acting in their own self-interest would produce what was best for the community as a whole (Smith 1776). Smith differed slightly from the physiocrats, however, on the idea of what led to wealth in a society. Smith believed that wealth sprang from the labor applied to the land (we might now say the "value added" to the land by labor), rather than it being just in the land itself. From this idea sprang the belief that there is a natural harmony between economic growth and population growth, with the latter depending always on the former. Thus, Smith felt that population size is determined by the demand for labor, which is, in turn, determined by the productivity of the land. These ideas are important to us because Smith's work served as an inspiration for the Malthusian theory of population, as Malthus himself acknowledges (see the preface to the sixth edition of Malthus 1872) and as I discuss later.

The Prelude to Malthus

The eighteenth century was the Age of Enlightenment in Europe, a time when the goodness of the common person was championed. This perspective, that the rights of individuals superseded the demands of a monarchy, inspired the American and French Revolutions and was generally very optimistic and utopian, characterized by a great deal of enthusiasm for life and a belief in the perfectibility of humans. It ushered in an era of critically questioning traditional ideas and authority that is still reverberating around the world.

In France, these ideas were well expressed by Marie Jean Antoine Nicolas de Caritat, marquis de Condorcet, a member of the French aristocracy who forsook a military career to pursue a life devoted to mathematics and philosophy. His ideas helped to shape the French Revolution, although despite his inspiration for and sympathy with that cause, he died in prison at the hands of revolutionaries. In hiding before his arrest, Condorcet wrote a *Sketch for an Historical Picture of the Progress of the Human Mind* (Condorcet 1795 [1955]). He was a visionary who "saw the outlines of liberal democracy more than a century in advance of his time: universal education; universal suffrage; equality before the law; freedom of thought and expression; the right to freedom and self-determination of colonial peoples; the redistribution of wealth; a system of national insurance and pensions; equal rights for women" (Hampshire 1955:x).

Condorcet's optimism was based on his belief that technological progress has no limits: "With all this progress in industry and welfare which establishes a happier proportion between men's talents and their needs, each successive generation will have larger possessions, either as a result of this progress or through the preservation of the products of industry, and so, *as a consequence of the physical constitution of the human race, the number of people will increase*" (Condorcet 1795 [1955]:188; emphasis added). He then asked whether it might not happen that eventually the happiness of the population would reach a limit. If that happens, Condorcet concluded, "we can assume that by then men will know that . . . their aim should be to promote the general welfare of the human race or of the society in which they live or of the family to which they belong, rather than foolishly to encumber the world with useless and wretched beings" (p. 189). Condorcet thus saw prosperity and population growth increasing hand in hand, and if the limits to growth were ever reached, the final solution would be birth control.

On the other side of the English Channel, similar ideas were being expressed by William Godwin (father of Mary Wollstonecraft Shelley, author of *Frankenstein*, and father-in-law of the poet Percy Bysshe Shelley). Godwin's *Enquiry Concerning Political Justice and Its Influences on Morals and Happiness* appeared in its first edition in 1793, revealing his ideas that scientific progress would enable the food supply to grow far beyond the levels of his day, and that such prosperity would not lead to overpopulation because people would deliberately limit their sexual expression and procreation. Furthermore, he believed that most of the problems of the poor were due not to overpopulation but to the inequities of the social institutions, especially greed and accumulation of property (Godwin 1793 [1946]).

Thomas Robert Malthus had recently graduated from Jesus College at Cambridge and was a country curate and a nonresident fellow of Cambridge as he read and contemplated the works of Godwin, Condorcet, and others who shared the utopian view of the perfectibility of human society. Although he wanted to be able to embrace such an openly optimistic philosophy of life, he felt that intellectually he had to reject it. In doing so, he unleashed a controversy about population growth and its consequences that rages to this very day.

The Malthusian Perspective

The **Malthusian** perspective derives from the writings of Thomas Robert Malthus ("Robert" to his family and friends), an English clergyman and subsequently a college professor. His first *Essay on the Principle of Population as it affects the future improvement of society; with remarks on the speculations of Mr. Godwin, M. Condorcet, and other writers* was published anonymously in 1798. Malthus's original intention was not to carve out a career in population studies, but only to show that the unbounded optimism of the physiocrats and utopian philosophers was misplaced. He introduced his essay by commenting that "I have read some of the speculations on the perfectibility of man and society, with great pleasure. I have been warmed and delighted with the enchanting picture which they hold forth. I ardently wish for such happy improvements. But I see great, and, to my understanding, unconquerable difficulties in the way to them" (Malthus 1798 [1965]:7).

These "difficulties," of course, are the problems posed by his now famous **principle of population**. He derived his theory as follows:

> I think I may fairly make two postulata. First, that food is necessary to the existence of man. Secondly, that the passion between the sexes is necessary, and will remain nearly in its present state. . . . Assuming then, my postulata as granted, I say, that the power of population is indefinitely greater than the power in the earth to produce subsistence for man. Population, when unchecked, increases in a geometrical ratio. Subsistence increases only in an arithmetical ratio. . . . By the law of our nature which makes food necessary to

> the life of man, the effects of these two unequal powers must be kept equal. This implies a strong and constantly operating check on population from the difficulty of subsistence.
>
> This difficulty must fall somewhere; and must necessarily be severely felt by a large portion of mankind. . . . Consequently, if the premises are just, the argument is conclusive against the perfectibility of the mass of mankind. (Malthus 1798 [1965]:11)

Malthus believed that he had demolished the utopian optimism by suggesting that the laws of nature, operating through the principle of population, essentially prescribed poverty for a certain segment of humanity. Malthus was a shy person by nature (James 1979; Petersen 1979), and he seemed ill prepared for the notoriety created by his essay. Nonetheless, after owning up to its authorship, he proceeded to document his population principles and to respond to critics by publishing a substantially revised version in 1803, slightly but importantly retitled to read *An Essay on the Principle of Population; or a view of its past and present effects on human happiness; with an inquiry into our prospects respecting the future removal or mitigation of the evils which It occasions.* In all, he published six editions of the book during his lifetime, followed by a seventh edition published posthumously (Malthus 1872 [1971]), and as a whole they have undoubtedly been the single most influential work relating population growth to its social consequences. Although Malthus initially relied on earlier writers such as David Hume (1752 [1963]) and Robert Wallace (1761 [1969]), he was the first to draw a picture that links the consequences of growth to its causes in a systematic way.

Causes of Population Growth

Malthus believed that human beings, like plants and nonrational animals, are "impelled" to increase the population of the species by what he called a powerful "instinct," the urge to reproduce. Further, if there were no checks on population growth, human beings would multiply to an "incalculable" number, filling "millions of worlds in a few thousand years" (Malthus 1872 [1971]:6). We humans, though, have not accomplished anything nearly so impressive. Why not? Because of the **checks to growth** that Malthus pointed out—factors that have kept population growth from reaching its biological potential for covering the earth with human bodies.

According to Malthus, the ultimate check to growth is lack of food (the **"means of subsistence"**). In turn, the means of subsistence are limited by the amount of land available, the "arts" or technology that could be applied to the land, and "social organization" or land ownership patterns. A cornerstone of his argument is that populations tend to grow more rapidly than the food supply does, since population has the potential for growing geometrically—two parents could have four children, sixteen grandchildren, and so on—while he believed (incorrectly, as Darwin later pointed out) that food production could be increased only arithmetically, by adding one acre at a time. This led to his conclusion that in the natural order of things, population growth will outstrip the food supply, and the lack of food will ultimately put a stop to the increase of people (see Figure 3.2).

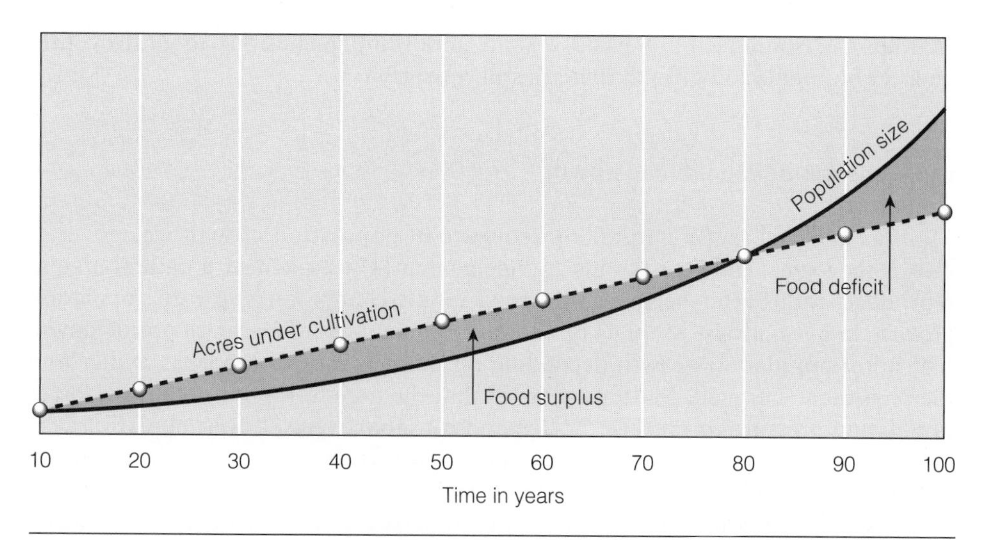

Figure 3.2 Over Time, Geometric Growth Overtakes Arithmetic Growth

Note: If we start with 100 acres supporting a population of 100 people and then add 100 acres of cultivated land per decade (arithmetic growth) while the population is increasing by 3 percent per year (geometric growth), the result is a few decades of food surplus before population growth overtakes the increase in the acres under cultivation, producing a food deficit, or "misery," as Malthus called it.

Of course, Malthus was aware that starvation rarely operates directly to kill people, since something else usually intervenes to kill them before they actually die of starvation. This "something else" represents what Malthus calls **positive checks**, primarily those measures "whether of a moral or physical nature, which tend prematurely to weaken and destroy the human frame" (Malthus 1872 [1971]:12). Today we would call these the causes of death. There are also **preventive checks**—limits to birth. In theory, the preventive checks would include all possible means of birth control, including abstinence, contraception, and abortion. However, to Malthus the only acceptable means of preventing a birth was to exercise **moral restraint**; that is, to postpone marriage, remaining chaste in the meantime, until a man feels "secure that, should he have a large family, his utmost exertions can save them from rags and squalid poverty, and their consequent degradation in the community" (1872 [1971]:13). Any other means of birth control, including contraception (either before or after marriage), abortion, infanticide, or any "improper means," was viewed as a vice that would "lower, in a marked manner, the dignity of human nature." Moral restraint was a very important point with Malthus, because he believed that if people were allowed to prevent births by "improper means" (that is, prostitution, contraception, abortion, or sterilization), then they would expend their energies in ways that are, so to speak, not economically productive.

As a scientific theory, the Malthusian perspective leaves much to be desired, since he was wrong about how quickly the food supply could increase, as I note below, and because he constantly confuses moralistic and scientific thinking (Davis 1955). Despite its shortcomings, however, which were evident even in his time,

Malthus's reasoning led him to draw some important conclusions about the consequences of population growth that are still relevant to us.

Consequences of Population Growth

Malthus believed that a natural consequence of population growth was poverty. This is the logical end result of his arguments that (1) people have a natural urge to reproduce, and (2) the increase in the food supply cannot keep up with population growth. In his analysis, Malthus turned the argument of Adam Smith upside down. Instead of population growth depending on the demand for labor, as Smith (and the physiocrats) argued, Malthus believed that the urge to reproduce always forces population pressure to precede the demand for labor. Thus, "overpopulation" (as measured by the level of unemployment) would force wages down to the point where people could not afford to marry and raise a family. At such low wages, with a surplus of labor and the need for each person to work harder just to earn a subsistence wage, cultivators could employ more labor, put more acres into production, and thus increase the means of subsistence. Malthus believed that this cycle of increased food resources, leading to population growth, leading to too many people for available resources, leading then back to poverty, was part of a natural law of population. Each increase in the food supply only meant that eventually more people would live in poverty.

As you can see, Malthus did not have an altogether high opinion of his fellow creatures. He figured that most of them were too "inert, sluggish, and averse from labor" (1798 [1965]:363) to try to harness the urge to reproduce and avoid the increase in numbers that would lead back to poverty whenever more resources were available. In this way, he essentially blamed poverty on the poor themselves. There remained only one improbable way to avoid this dreary situation.

Avoiding the Consequences

Borrowing from John Locke, Malthus argued that "the endeavor to avoid pain rather than the pursuit of pleasure is the great stimulus to action in life" (1798 [1965]:359). Pleasure will not stimulate activity until its absence is defined as being painful. Malthus suggested that the well-educated, rational person would perceive in advance the pain of having hungry children or being in debt and would postpone marriage and sexual intercourse until he was sure that he could avoid that pain. If that motivation existed and the preventive check was operating, then the miserable consequences of population growth could be avoided. You will recall that Condorcet had suggested the possibility of birth control as a preventive check, but Malthus objected to this solution: "To remove the difficulty in this way, will, surely in the opinion of most men, be to destroy that virtue, and purity of manners, which the advocates of equality, and of the perfectibility of man, profess to be the end and object of their views" (1798:154). So the only way to break the cycle is to change human nature. Malthus felt that if everyone shared middle-class values, the problem would solve itself. He saw that as impossible, though, since not everyone has the talent to be a virtuous, industrious, middle-class success story, but if most people at least tried, poverty would be reduced considerably.

To Malthus, material success is a consequence of the human ability to plan rationally—to be educated about future consequences of current behavior—and he was a man who practiced what he preached. He planned his family rationally, waiting to marry and have children until he was 39, shortly after getting a secure job in 1805 as a professor of history and political economy at East India College in Haileybury, England (north of London). Also, although Marx thought that Malthus had taken the "monastic vows of celibacy" whereas other detractors attributed 11 children to him, Malthus and his wife, 11 years his junior, had only three children (Nickerson 1975; Petersen 1979).

To summarize, the major consequence of population growth, according to Malthus, is poverty. Within that poverty, though, is the stimulus for action that can lift people out of misery. So, if people remain poor, it is their own fault for not trying to do something about it. For that reason, Malthus was opposed to the English Poor Laws (welfare benefits for the poor), because he felt they would actually serve to perpetuate misery. They permitted poor people to be supported by others and thus not feel that great pain, the avoidance of which might lead to birth prevention. Malthus argued that if every man had to provide for his own children, he would be more prudent about getting married and raising a family. In his own time, this particular conclusion of Malthus brought him the greatest notoriety, because the number of people on welfare had been increasing and English parliamentarians were trying to decide what to do about the problem. Although the Poor Laws were not abolished, they were reformed largely because Malthus had given legitimacy to public criticism of the entire concept of welfare payments (Himmelfarb 1984). The Malthusian perspective that blamed the poor for their own poverty endures, contrasted with the equally enduring view of Godwin and Condorcet that poverty is the creation of unjust human institutions. Two hundred additional years of debate have only sharpened the edges of the controversy.

Critique of Malthus

The single most obvious measure of Malthus's importance is the number of books and articles that have attacked him, beginning virtually the moment his first essay appeared in 1798 and continuing to the present (see, for example, Lee and Wang Feng 1999; Huzel 2006; Sabin 2013). Hodgson (2009) quotes from a letter written by Thomas Jefferson in 1804 discussing the fact that he had just read Malthus's book and that he (Jefferson) was sure the principle of population did not apply to the United States, where the amount of available land meant that population growth could readily be absorbed. But Hodgson notes that later in the nineteenth century both sides in the debate over ending slavery in the United States called upon Malthusian arguments to bolster their case, even though Malthus himself was vociferously opposed to slavery.

The three most strongly criticized aspects of his theory have been (1) the assertion that food production could not keep up with population growth, (2) the conclusion that poverty was an inevitable result of population growth, and (3) the belief that moral restraint was the only acceptable preventive check. Malthus was not a firm believer in progress; rather, he accepted the notion that each society had a fixed set of institutions that established a stationary level of living. He was aware, of course, of the Industrial Revolution, but he was skeptical of its long-run value and agreed with the physiocrats that real wealth was in agricultural land. He was convinced that the increase in manufacturing wages that accompanied industrialization would promote population growth without increasing the agricultural production necessary to feed those additional mouths. Although it is clear that he was a voracious reader (Petersen 1999) and was a founder of the Statistical Society of London (Starr 1987), it is also clear that Malthus paid scant attention to the economic statistics that were available to him. "There is no sign that even at the end of his life he knew anything in detail about industrialization. His thesis was based on the life of an island agricultural nation, and so it remained long after the exports of manufacturers had begun to pay for the imports of large quantities of raw materials" (Eversley 1959:256). Thus, Malthus either failed to see or refused to acknowledge that technological progress was possible, and that its end result was a higher standard of living, not a lower one.

The crucial part of Malthus's ratio of population growth to food increase was that food (including both plants and nonhuman animals) would not grow exponentially, whereas humans could grow like that. Yet when Charles Darwin acknowledged that his concept of the survival of the fittest was inspired by Malthus's essay, he implicitly rejected this central tenet of Malthus's argument. In Chapter Three of *On the Origin of Species*, Darwin described his own theory as "the doctrine of Malthus applied with manifold force to the whole animal and vegetable kingdoms; for in this case there can be no artificial increase of food, and no prudential restraint from marriage. Although some species may be now increasing, more or less rapidly, in numbers, all cannot do so, for the world would not hold them" (Darwin 1872 [1991]:47).

Malthus's argument that poverty is an inevitable result of population growth is also open to scrutiny. For one thing, his writing reveals a certain circularity in logic. In Malthus's view, a laborer could achieve a higher standard of living only by being prudent and refraining from marriage until he could afford it, but Malthus also believed that you could not expect prudence from a laborer until he had attained a higher standard of living. Thus, our hypothetical laborer seems squarely enmeshed in a catch-22. Even if we were to ignore this logical inconsistency, there are problems with Malthus's belief that the Poor Laws contributed to the misery of the poor by discouraging them from exercising prudence. Historical evidence has revealed that between 1801 and 1835 those English parishes that administered Poor Law allowances did not have higher birth, marriage, or total population growth rates than those in which Poor Law assistance was not available (Huzel 1969, 1980, 1984). Clearly, problems with the logic of Malthus's argument seem to be compounded by his apparent inability to see the social world accurately: "The results of the 1831 Census were out before he died, yet he never came to interpret them. Statistics apart, the main charge against him must be that he was a bad observer of his fellow human beings" (Eversley 1959:256).

I noted in Chapter 1 that the term "demography" was first used by a French scientist, Achille Guillard, in the middle of the nineteenth century. Schweber (2006) has argued that one of Guillard's motivations in trying to develop a new discipline of demography was to pressure French academics to see that statistical analyses of births and deaths would show that Malthus was wrong about his claim that population growth inevitably led to poverty. Once again, the power of Malthusian thought lies partly in the strength of opposition that he aroused.

Neo-Malthusians

Those who criticize Malthus's insistence on the value of moral restraint, while accepting many of his other conclusions, are typically known as **neo-Malthusians** (see the essay in this chapter for more discussion). Specifically, neo-Malthusians favor contraception rather than simple reliance on moral restraint. During his lifetime, Malthus was constantly defending moral restraint against critics (many of whom were his friends) who encouraged him to deal more favorably with other means of birth control. In the fifth edition of his *Essay*, he did discuss the concept of *prudential restraint*, which meant the delay of marriage until a family could be afforded without necessarily refraining from premarital sexual intercourse in the meantime. He never fully embraced the idea, however, nor did he ever bow to pressure to accept anything but moral restraint as a viable preventive check.

Ironically, the open controversy actually helped to spread knowledge of birth control among people in nineteenth-century England and America. This was aided materially by the trial and conviction (later overturned on a technicality) in 1877–78 of two neo-Malthusians, Charles Bradlaugh and Annie Besant, for publishing a birth control handbook (*Fruits of Philosophy: The Private Companion of Young Married People*, written by Charles Knowlton, a physician in Massachusetts, and originally published in 1832). The publicity surrounding the trial enabled the English public to become more widely knowledgeable about those techniques (Chandrasekhar 1979). Eventually, the widespread adoption of birth control meant that fertility could be controlled *within* marriage, allowing couples to respond to economic changes in ways that were not anticipated by Malthus's principle of population.

Criticisms of Malthus do not, however, diminish the importance of his work:

> There are good reasons for using Malthus as a point of departure in the discussion of population theory. These are the reasons that made his work influential in his day and make it influential now. But they have little to do with whether his views are right or wrong. . . . Malthus' theories are not now and never were empirically valid, but they nevertheless were theoretically significant. (Davis 1955b:541)

As I noted earlier, part of Malthus's significance lies in the storm of controversy his theories stimulated. Particularly vigorous in their attacks on Malthus were Karl Marx and Friedrich Engels.

WHO ARE THE NEO-MALTHUSIANS?

"Picture a tropical island with luscious breadfruits [a Polynesian plant similar to a fig tree] hanging from every branch, toasting in the sun. It is a small island, but there are only 400 of us on it so there are more breadfruits than we know what to do with. We're rich. Now picture 4,000 people on the same island, reaching for the same breadfruits: Number one, there are fewer to go around; number two, you've got to build ladders to reach most of them; number three, the island is becoming littered with breadfruit crumbs. Things get worse and worse as the population gradually expands to 40,000. Welcome to a poor, littered tropical paradise" (Tobias 1979:49). This scenario would probably have drawn a nod of understanding from Malthus himself, and even though written a few decades ago, it typifies the modern neo-Malthusian view of the world.

One of the most influential neo-Malthusians of the twentieth century was the University of California, Santa Barbara biologist Garrett Hardin. In 1968, he published an article that raised the level of consciousness about population growth in the minds of professional scientists. Hardin's theme was simple and had been made by Kingsley Davis (1963) as he developed the theory of demographic change and response: Personal goals are not necessarily consistent with societal goals when it comes to population growth—Adam Smith was not completely correct to believe in laissez-faire. Hardin's metaphor is "the tragedy of the commons." He asks us to imagine an open field, available as a common ground for herdsmen to graze their cattle: "As a rational being, each herdsman seeks to maximize his gain. Explicitly or implicitly, more or less consciously, he asks, 'What is the utility to me of adding one more animal to my herd?'" (Hardin 1968:1244). The benefit, of course, is the net proceeds from the eventual sale of each additional animal, whereas the cost lies in the chance that an additional animal may result in overgrazing of the common ground. Since the ground is shared by many people, the cost is spread out over all, so for the individual herdsman, the benefit of another animal exceeds its cost. "But," notes Hardin, "this is the conclusion reached by each and every rational herdsman sharing a commons. Therein is the tragedy. Each man is locked into a system that compels him to increase his herd without limit—in a world that is limited" (1968:1244). The moral, as Hardin puts it, is that "ruin is the destination toward which all men rush, each pursuing his own best interest in a society that believes in the freedom of the commons. Freedom in a commons brings ruin to all" (1968:1244).

Hardin reminds us that most societies are committed to a social welfare ideal. Families are not completely on their own. We share numerous things in common: education, public health, and police protection, and in all of the richer nations of the world people are guaranteed a minimum amount of food and income at the public expense. This leads to a moral dilemma that is at the heart of Hardin's message: "To couple the concept of freedom to breed with the belief that everyone born has an equal right to the commons is to lock the world into a tragic course of action" (Hardin 1968:1246). He was referring, of course, to the ultimate Malthusian clash of population and resources, and Hardin was no more optimistic than Malthus about the likelihood of people voluntarily limiting their fertility before it is too late.

Meanwhile, in the 1960s the world was becoming keenly aware of the population crisis through the writings of the person who is arguably the most famous of all neo-Malthusians, Paul Ehrlich. Like Hardin, Ehrlich is a biologist (at Stanford University), not a professional demographer. His *Population Bomb* (Ehrlich 1968) was an immediate sensation when it came out in 1968 and to this day often sets the tone for public debate about population issues. In the second edition of his book, Ehrlich (1971) phrased the situation in three parts: "too many people," "too little food," and, adding a wrinkle not foreseen directly by Malthus, "environmental degradation" (Ehrlich called earth "a dying planet").

In 1990, Ehrlich, in collaboration with his wife, Anne, followed with an update titled The *Population Explosion* (Ehrlich and Ehrlich 1990), reflecting their view that the bomb they worried about in 1968 had detonated in the meantime. The level of concern about the destruction of the environment has grown tremendously since 1968. Ehrlich's book had inspired the first Earth Day in the spring of 1970 (an annual event ever since in most communities across the United States and elsewhere in the world), yet in their 1990 book Ehrlich and Ehrlich rightly question why, in the face of the serious environmental degradation that had concerned them for so long, had people regularly failed to grasp its primary cause as being rapid population growth? "Arresting population growth should be second in importance only to avoiding nuclear war on humanity's agenda. Overpopulation and rapid population growth are intimately connected with most aspects of the current human predicament, including rapid depletion of nonrenewable resources, deterioration of the environment (including rapid climate change), and increasing international tensions" (Ehrlich and Ehrlich 1990:18).

Ehrlich thus argues that Malthus was right—dead right. But the death struggle is more complicated than that foreseen by Malthus. To Ehrlich, the poor are dying of hunger, while rich and poor alike are dying from the by-products of affluence—pollution and ecological disaster. Indeed, this is part of the "commons" problem. A few benefit; all suffer. What does the future hold? Ehrlich suggested that there are only two solutions to the population problem: the birth rate solution (lowering the birth rate) and the death rate solution (a rise in the death rate). He viewed the death rate solution as being the most likely to happen, because, like Malthus, he has had little faith in the ability of humankind to pull its act together. The only way to avoid that scenario, he argued, was to bring the birth rate under control, perhaps even by force. That idea generated death threats against him and his wife, but of course over time the world has responded with lower birth rates.

A major part of Ehrlich's contribution has been to encourage people to take some action themselves, to spread the word and practice what they preach. Ehrlich has long felt that population growth is outstripping resources and ruining the environment. If we sit back and wait for people to react to this situation, disaster will occur. Therefore, we need to act swiftly to push people to bring fertility down to replacement level by whatever means possible.

Neo-Malthusians thus differ from Malthus because they reject moral restraint as the only acceptable means of birth control and because they see population growth as leading not simply to poverty but also to widespread calamity. For neo-Malthusians, the "evil arising from the redundancy of population" that Malthus worried about has broadened in scope, and the remedies proposed are thus more dramatic.

Gloomy they certainly are, but the messages of Ehrlich and Hardin are important and impressive and have brought population issues to the attention of the entire globe. One of the ironies of neo-Malthusianism is that if the world's population does avoid future calamity, people will likely claim that the neo-Malthusians were wrong. In fact, however, much of the stimulus to bring down birth rates (including emphasis on the reproductive rights of women as the alternative to coercive means) and to find new ways to feed people and protect the environment has come as a reaction to the concerns they very publicly have raised.

Discussion Questions: (1) Discuss the tragedy of the commons in relation to global climate change and to the quality of water throughout the world, and relate that to population growth; **(2)** If Malthus was wrong in his idea that the food supply could not grow as quickly as population, as Darwin seemed to suggest, do you think that the neo-Malthusians are also wrong in their analysis of how the world works? Why or why not?

The Marxian Perspective

Karl Marx and Friedrich Engels were both teenagers in Germany when Malthus died in England in 1834, and by the time they had met and independently moved to England, Malthus's ideas already were politically influential in their native land, not just in England. Several German states and Austria had responded to what they believed was overly rapid growth in the number of poor people by legislating against marriages in which the applicant could not guarantee that his family would not wind up on welfare (Glass 1953). As it turned out, that scheme backfired on the German states, because people continued to have children, but out of wedlock. Thus, the welfare rolls grew as the illegitimate children had to be cared for by the state (Knodel 1970). The laws were eventually repealed, but they had an impact on Marx and Engels, who saw the Malthusian point of view as an outrage against humanity. Their demographic perspective thus arose in reaction to Malthus.

Causes of Population Growth

Neither Marx nor Engels ever directly addressed the issue of why and how populations grew. They seem to have had little quarrel with Malthus on this point, although they were in favor of equal rights for men and women and saw no harm in preventing birth. Nonetheless, they were skeptical of the eternal or natural laws of nature as stated by Malthus (that population tends to outstrip resources), preferring instead to view human activity as the product of a particular social and economic environment. The basic **Marxian** perspective is that each society at each point in history has its own law of population that determines the consequences of population growth. For **capitalism**, the consequences are overpopulation and poverty, whereas for **socialism**, population growth is readily absorbed by the economy with no side effects. This line of reasoning led to Marx's vehement rejection of Malthus, because if Malthus was right about his "pretended 'natural law of population'" (Marx 1890 [1906]:680), then Marx's theory would be wrong.

Consequences of Population Growth

Marx and Engels especially quarreled with the Malthusian idea that resources could not grow as rapidly as population, since they saw no reason to suspect that science and technology could not increase the availability of food and other goods at least as quickly as the population grew. Engels argued in 1865 that whatever population pressure existed in society was really pressure against the means of employment rather than against the means of subsistence (Meek 1971). Thus, they flatly rejected the notion that poverty can be blamed on the poor. Instead, they said, poverty is the result of a poorly organized society, especially a capitalist society. Implicit in the writings of Marx and Engels is the idea that the normal consequence of population growth should be a significant increase in production. After all, each worker obviously was producing more than he or she required—how else would all the dependents (including the wealthy manufacturers) survive? In a well-ordered society, if there were more people, there ought to be more wealth, not more poverty (Engels 1844 [1953]).

Not only did Marx and Engels feel that poverty, in general, was not the end result of population growth, they argued specifically that even in England at that time there was enough wealth to eliminate poverty. Engels had himself managed a textile plant owned by his father's firm in Manchester, and he believed that in England more people had meant more wealth for the capitalists rather than for the workers because the capitalists were skimming off some of the workers' wages as profits for themselves. Marx argued that they did that by stripping the workers of their tools and then, in essence, charging the workers for being able to come to the factory to work. For example, if you do not have the tools to make a car but want a job making cars, you could get hired at the factory and work eight hours a day. But, according to Marx, you might get paid for only four hours, the capitalist (owner of the factory) keeping part of your wages as payment for the tools you were using. The more the capitalist keeps, of course, the lower your wages and the poorer you will be.

Furthermore, Marx argued that capitalism worked by using the labor of the working classes to earn profits to buy machines that would replace the laborers, which, in turn, would lead to unemployment and poverty. Thus, the poor were not poor because they overran the food supply, but only because capitalists had first taken away part of their wages and then taken away their very jobs and replaced them with machines. Thus, the consequences of population growth that Malthus discussed were really the consequences of capitalist society, not of population growth per se. Overpopulation in a capitalist society was thought to be a result of the capitalists' desire for an industrial reserve army that would keep wages low through competition for jobs and, at the same time, would force workers to be more productive in order to keep their jobs. To Marx, the logical extension of this was that the growing population would bear the seeds of destruction for capitalism, because unemployment would lead to disaffection and revolution. If society could be reorganized in a more equitable (that is, socialist) way, then population problems would disappear.

It is noteworthy that Marx, like Malthus, practiced what he preached. Marx was adamantly opposed to the notion of moral restraint, and his life repudiated that concept. He married at the relatively young age (compared with Malthus) of 25, proceeded to father eight children, including one illegitimate son, and was on intimate terms with poverty for much of his life.

In its original formulation, the Marxian (as well as the Malthusian) perspective was somewhat provincial, in the sense that its primary concern was England in the nineteenth century. Marx was an intense scholar who focused especially on the historical analysis of economics as applied to England, which he considered to be the classic example of capitalism. However, as his writings have found favor in other places and times, revisions have been forced upon the Marxian view of population.

Critique of Marx

Not all who have adopted a Marxian worldview fully share the original Marx–Engels demographic perspective. Socialist countries have had trouble because of the lack of political direction offered by the Marxian notion that different stages of social development produce different relationships between population growth and economic development. Indeed, much of what we call the Marxian thought on population is in fact attributable to Lenin, one of the most prolific interpreters of Marx. For Marx, the Malthusian principle operated under capitalism only, whereas under pure socialism there would be no population problem. Unfortunately, he offered no guidelines for the transition period. At best, Marx implied that the socialist law of population should be the antithesis of the capitalist law. If the birth rate were low under capitalism, then the assumption was that it should be high under socialism; if abortion seemed bad for a capitalist society, it must be good for a socialistic society.

Thus, it was difficult for Russian demographers to reconcile the fact that demographic trends in the former Soviet Union were remarkably similar to trends in other developed nations. Furthermore, Soviet socialism was unable to alleviate one of the worst evils that Marx attributed to capitalism, higher death rates among people in the working class than among those in the higher classes (Brackett 1968). Moreover, birth rates dropped to such low levels throughout Marxist Eastern Europe in the years leading up to the breakup of the Soviet Union that it was no longer possible to claim (as Marx had done) that low birth rates were bourgeois.

In China, the empirical reality of having to deal with the world's largest national population led to a radical departure from Marxian ideology. As early as 1953, the Chinese government organized efforts to control population by relaxing regulations concerning contraception and abortion. Ironically, after the terrible demographic disaster that followed the "Great Leap Forward" in 1958 (see Chapter 2), a Chinese official quoted Chairman Mao as having said, "A large population in China is a good thing. With a population increase of several fold we still have an adequate solution. The solution lies in production" (Ta-k'un 1960:704). Yet by the 1970s production no longer seemed to be a panacea, and with the introduction of the one-child policy in 1979, the interpretation of Marx took an about-face as another Chinese official wrote that under Marxism the law of production "demands not only a planned production of natural goods, but also the planned reproduction of human beings" (Muhua 1979:724).

Thus, despite Marx's denial of a population problem in a socialist society, the Marxist government in China dealt with one by rejecting its Marxist–Leninist roots and embracing instead one of the most aggressive and coercive government programs ever launched to reduce fertility through restraints on marriage (the Malthusian solution), the promotion of contraception (the neo-Malthusian solution), and the use of abortion (a remnant of the Leninist approach) (Teitelbaum and Winter 1988). In a formulation such as this, Marxism was revised in the light of new scientific evidence about how people behave, in the same way that Malthusian thought has been revised. Bear in mind that although the Marxian and Malthusian perspectives are often seen as antithetical, they both originated in the midst of a particular milieu of economic, social, and demographic change in nineteenth-century Europe.

The Prelude to the Demographic Transition Theory

The population-growth controversy, initiated by Malthus and fueled by Marx, emerged into a series of nineteenth-century and early-twentieth-century reformulations that have led directly to prevailing theories in demography. In this section, I briefly discuss three individuals who made important contributions to those reformulations: John Stuart Mill, Arsène Dumont, and Émile Durkheim.

Mill

The English philosopher and economist John Stuart Mill was an extremely influential writer of the nineteenth century. Mill was not as quarrelsome about Malthus as Marx and Engels had been; his scientific insights were greater than those of Malthus at the same time that his politics were less radical than those of Marx and Engels. Mill accepted the Malthusian calculations about the potential for population growth to outstrip food production as being axiomatic (a self-truth), but he was more optimistic about human nature than Malthus was. Mill believed that although your character is formed by circumstances, one's own desires can do much to shape circumstances and modify future habits (Mill 1873 [1924]).

Mill's basic thesis was that the standard of living is a major determinant of fertility levels: "In proportion as mankind rises above the condition of the beast, population is restrained by the fear of want, rather than by want itself. Even where there is no question of starvation, many are similarly acted upon by the apprehension of losing what have come to be regarded as the decencies of their situation in life" (Mill 1848 [1929]:Book I, Chap 10). The belief that people could be and should be free to pursue their own goals in life led him to reject the idea that poverty is inevitable (as Malthus implied) or that it is the creation of capitalist society (as Marx argued). One of Mill's most famous comments is that "the niggardliness of nature, not the injustice of society, is the cause of the penalty attached to overpopulation" (1848 [1929]:Book I, Chap. 13). This is a point of view conditioned by Mill's reading of Malthus, but Mill denies the Malthusian inevitability of a population growing beyond its available resources. Mill believed that people do not "propagate like swine, but are capable, though in very unequal degrees, of being withheld by prudence, or by the social affections, from giving existence to beings born only to misery and premature death" (1848 [1929]: Book I, Chap. 7). In the event that population ever did overrun the food supply, however, Mill felt that it would likely be a temporary situation with at least two possible solutions: import food or export people.

The ideal state from Mill's point of view is that in which all members of a society are economically comfortable. At that point he felt (as Plato had centuries earlier) that the population would stabilize and people would try to progress culturally, morally, and socially instead of attempting continually to get ahead economically. It does sound good, but how do we get to that point? It was Mill's belief that before reaching the point at which both population and production are stable, there is essentially a race between the two. What is required to settle the issue is a dramatic improvement in the living conditions of the poor. If social and economic development are to occur, there must be a sudden increase in income, which could give rise to a new standard of living for a whole generation, thus allowing productivity to outdistance population growth. According to Mill, this was the situation in France after the Revolution:

> During the generation which the Revolution raised from the extremes of hopeless wretchedness to sudden abundance, a great increase of population took place. But a generation has grown up, which, having been born in improved circumstances, has not learnt to be miserable; and upon them the spirit of thrift operates most conspicuously, in keeping the increase of population within the increase of national wealth. (1848 [1929]: Book II, Chap. 7)

Mill was convinced that an important ingredient in the transformation to a non-growing population is that women do not want as many children as men do, and if they are allowed to voice their opinions, the birth rate will decline. Mill, like Marx, was a champion of equal rights for both sexes, and one of Mill's more notable essays, *On Liberty*, was co-authored with his wife. He reasoned further that a system of national education for poor children would provide them with the "common sense" (as Mill put it) to refrain from having too many children.

Overall, Mill's perspective on population growth was significant enough that we find his arguments surviving today in the writings of many of the twentieth- and twenty-first-century demographers whose names appear in the pages that follow. However, before getting to those people and their ideas, it is important to acknowledge at least two other nineteenth-century individuals whose thinking has an amazingly modern sound: Arsène Dumont and Émile Durkheim.

Dumont

Arsène Dumont was a late-nineteenth-century French demographer who felt he had discovered a new principle of population that he called "social capillarity" (Dumont 1890). **Social capillarity** refers to the desire of people to rise on the social scale, to increase their individuality as well as their personal wealth. The concept is drawn from an analogy to a liquid rising into the narrow neck of a laboratory flask. The flask is like the hierarchical structure of most societies, broad at the bottom and narrowing as you near the top. To ascend the social hierarchy often requires that sacrifices be made, and Dumont argued that having few or no children was the price many people paid to get ahead. Dumont recognized that such ambitions were not possible in every society. In a highly stratified aristocracy, few people outside of the aristocracy could aspire to a career beyond subsistence. However, in a democracy (such as late-nineteenth-century France), opportunities to succeed existed at all social levels. Spengler (1979) has succinctly summarized Dumont's thesis: "The bulk of the population, therefore, not only strove to ascend politically, economically, socially, and intellectually, but experienced an imperative urge to climb and a palsying fear of descent. Consequently, since children impeded individual and familial ascension, their number was limited" (p. 158).

Notice that Dumont added an important ingredient to Mill's recipe for fertility control. Mill argued that it was fear of social slippage that motivated people to limit fertility below the level that Malthus had expected. Dumont went beyond that to suggest that social aspiration was a root cause of a slowdown in population growth. Dumont was not happy with this situation, by the way. He was upset by the low level of French fertility and used the concept of social capillarity to propose policies to undermine it. He believed that socialism would undercut the desire for upward social mobility and would thus stimulate the birth rate. History, of course, has suggested otherwise, which would lead us back to the importance of Mill's view of the world.

Durkheim

While Dumont was concerned primarily with the causes of population growth, focusing mainly on the birth rate, another late-nineteenth-century French sociologist, Émile Durkheim, based an entire social theory on the consequences of population growth. In discussing the increasing complexity of modern societies, characterized particularly by increasing divisions of labor, Durkheim proposed that "the division of labor varies in direct ratio with the volume and density of societies, and, if it progresses in a continuous manner in the course of social development, it is because societies become regularly denser and more voluminous" (Durkheim 1893 [1933]:262). Durkheim proceeded to explain that population growth leads to greater societal specialization because the struggle for existence is more acute when there are more people. If you compare a primitive society with an industrialized society, the primitive society is not very specialized. By contrast, in industrialized societies there is a lot of differentiation; that is, there is an increasingly long list of occupations and social classes. Why is this? The answer is in the volume and density of the population. Growth creates competition for society's resources, and in order to improve their advantage in the struggle, people specialize.

Durkheim's thesis that population growth leads to specialization was derived (he himself acknowledged) from Darwin's theory of evolution. In turn, Darwin acknowledged his own debt to Malthus. You will notice that Durkheim also clearly echoes the words of Ibn Khaldun, although it is uncertain whether Durkheim knew of the latter's work.

The critical theorizing of the nineteenth and early twentieth centuries set the stage for a more systematic collection of data to test aspects of those theories and to examine more carefully those that might be valid and those that should be discarded. As population studies became more quantitative in the twentieth century, a phenomenon called the demographic transition took shape and took the attention of demographers.

The Theory of the Demographic Transition

Although it has dominated demographic thinking for the past half century, and is now almost routinely included in introductory texts in the social and environmental sciences, the **demographic transition** theory actually began as only a description of the demographic changes that had taken place in the advanced nations over time. In particular, it described the transition from high birth and death rates to low birth and death rates, with an interstitial spurt in growth rates leading to a larger population at the end of the transition than there had been at the start. The idea emerged as early as 1929, when Warren Thompson gathered data from "certain countries" for the period 1908–27 and showed that the countries fell into three main groups, according to their patterns of population growth:

> *Group A (northern and western Europe and the United States):* From the latter part of the nineteenth century to 1927, these countries had moved from having very high rates of natural increase to having very low rates of increase "and will shortly become stationary and start to decline in numbers." (Thompson 1929:968)
>
> *Group B (Italy, Spain, and the "Slavic" peoples of central Europe):* Thompson saw evidence of a decline in both birth rates and death rates but suggested that "it appears probable that the death rate will decline as rapidly or even more rapidly than the birth rate for some time yet. The condition in these Group B countries is much the same as existed in the Group A countries thirty to fifty years ago." (p. 968)
>
> *Group C (the rest of the world):* In the rest of the world, Thompson saw little evidence of control over either births or deaths.

As a consequence of this relative lack of voluntary control over births and deaths (a concept we will question later), Thompson felt that the Group C countries (which included about 70–75 percent of the population of the world at the time) would continue to have their growth "determined largely by the opportunities they have to increase their means of subsistence. Malthus described their processes of growth quite accurately when he said 'that population does invariably increase, where there are means of subsistence. . . .'" (Thompson, 1929:971).

Thompson's work, however, came at a time when there was little concern about overpopulation. The "Group C" countries had relatively low rates of growth because of high mortality and, at the same time, by 1936, birth rates in the United States and Europe were so low that Enid Charles published a widely read book called *The Twilight of Parenthood*, which was introduced with the comment that "in place of the Malthusian menace of overpopulation there is now real danger of underpopulation" (Charles 1936:v). Furthermore, Thompson's labels for his categories had little charisma. It is difficult to build a compelling theory around categories called A, B, and C.

Sixteen years after Thompson's work, Frank Notestein (1945) picked up the threads of his thesis and provided labels for the three types of growth patterns that Thompson had simply called A, B, and C. Notestein called the Group A pattern **incipient decline**, the Group B pattern **transitional growth**, and the Group C pattern **high growth potential**. That same year, Kingsley Davis (1945) edited a volume of the *Annals of the American Academy of Political and Social Sciences* titled "World Population in Transition," and in the lead article (titled "The Demographic Transition") he noted that "viewed in the long-run, earth's population has been like a long, thin powder fuse that burns slowly and haltingly until it finally reaches the charge and explodes" (Davis 1945:1). The term population explosion, alluded to by Davis, refers to the phase that Notestein called transitional growth. Thus was born the term *demographic transition*.

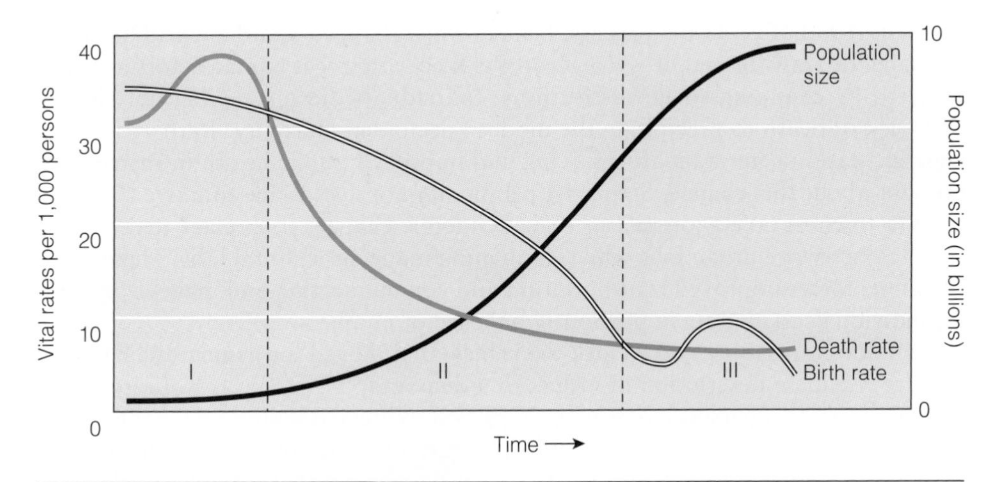

Figure 3.3 The Demographic Transition

Note: The original model of the demographic transition is divided roughly into three stages. In the first stage there is high growth potential because both birth and death rates are high. The second stage is the transition from high to low birth and death rates. During this stage the growth potential is realized as the death rate drops before the birth rate drops, resulting in rapid population growth. Finally, the last stage is a time when death rates are as low as they are likely to go, while fertility may continue to decline to the point that the population might eventually decline in numbers.

It is that process of moving from high birth and death rates to low birth and death rates, from high growth potential to incipient decline (see Figure 3.3).

At this point in the 1940s, however, the demographic transition was merely a picture of demographic change, not a theory. But each new country studied fit into the picture, and it seemed as though some new universal law of population growth—an evolutionary scheme—was being developed. The apparent historical uniqueness of the demographic transition (all known cases have occurred within the last 200 years) has spawned a host of alternative names, such as the "vital revolution" and the "demographic revolution." Between the mid-1940s and the late 1960s, rapid population growth became a worldwide concern, and demographers devoted a great deal of time to the demographic transition perspective. By 1964, George Stolnitz was able to report that "demographic transitions rank among the most sweeping and best-documented trends of modern times . . . based upon hundreds of investigations, covering a host of specific places, periods and events" (Stolnitz 1964:20). As the pattern of change took shape, explanations were developed for why and how countries pass through the transition. These explanations tended to be cobbled together in a somewhat piecemeal fashion from the nineteenth- and early-twentieth-century writers I discussed earlier in this chapter, but overall they were derived from the concept of **modernization**.

Modernization theory is based on the idea that in premodern times human society was generally governed by "tradition," and that the massive economic changes wrought by industrialization forced societies to alter traditional institutions: "In traditional societies fertility and mortality are high. In modern societies fertility and mortality are low. In between, there is demographic transition"

(Demeny 1968:502). In the process, behavior has changed and the world has been permanently transformed. It is a macro-level theory that sees human actors as being buffeted by changing social institutions. Individuals did not deliberately lower their risk of death to precipitate the modern decline in mortality. Rather, society-wide increases in the standard of living and improved public health infrastructure brought about this change. Similarly, people did not just decide to move from the farm to town to take a job in a factory. Economic changes took place that created those higher-wage urban jobs while eliminating many agricultural jobs. These same economic forces improved transportation and communication and made it possible for individuals to migrate in previously unheard of numbers.

Modernization theory provided the vehicle that moved the demographic transition from a mere description of events to a demographic perspective. In its initial formulations, this perspective was perhaps best expressed by the sentiments "take care of the people and population will take care of itself" or "development is the best contraceptive" (Teitelbaum 1975). These were views that were derivable from Karl Marx, who was in fact one of the early exponents of the modernization theory (Inglehart and Baker 2000). The theory drew on the available data for most countries that had gone through the transition. Death rates declined as the standard of living improved, and birth rates almost always declined a few decades later, eventually dropping to low levels, although rarely as low as the death rate. It was argued that the decline in the birth rate typically lagged behind the decline in the death rate because it takes time for a population to adjust to the fact that mortality really is lower, and because the social and economic institutions that favored high fertility require time to adjust to new norms of lower fertility that are more consistent with the lower levels of mortality. Since most people value the prolongation of life, it is not hard to lower mortality, but the reduction of fertility is contrary to the established norms of societies that have required high birth rates to keep pace with high death rates. Such norms are not easily changed, even in the face of poverty.

Birth rates eventually declined, it was argued, as the importance of family life was diminished by industrial and urban life, thus weakening the pressure for large families. Large families are presumed to have been desired because they provided parents with a built-in labor pool, and because children provided old-age security for parents. The same economic development that lowered mortality is theorized to transform a society into an urban industrial state in which compulsory education lowers the value of children by removing them from the labor force, and people come to realize that lower infant mortality means that fewer children need to be born to achieve a certain number of surviving children. Finally, as a consequence of the many alterations in social institutions, "the pressures for high fertility weaken and the idea of conscious control of fertility gradually gains strength" (Teitelbaum 1975:421).

Critique of the Demographic Transition Theory

It has been argued that the concept underlying the demographic transition is that population stability, also known as **homeostasis** (Lee 1987) is the normal state of affairs in human societies and that change (the "transition") is what requires explanation (Kreager 1986). Not everyone agrees. Harbison and Robinson (2002) argue that transitions are the natural state of human affairs, and that each transition is followed by another one, a theme we will return to later in the chapter. In its original formulation, the demographic transition theory explained high fertility as a reaction to high mortality. As mortality declines, the need for high fertility lessens, and so birth rates go down. There is a spurt of growth in that transition period, but presumably the consequences will not be serious if the decline in mortality was produced by a rise in the standard of living, which, in its turn, produces a motivation for smaller families. But what will be the consequences if mortality declines and fertility does not? That situation presumably is precluded by the theory of demographic transition, but the demographic transition theory has not been capable of predicting levels of mortality or fertility or the timing of the fertility decline. This is because the initial explanation for the demographic behavior during the transition tended to be **ethnocentric**. It relied almost exclusively on the sentiment that "what is good for the goose is good for the gander." In other words, if this is what happened to the developed countries, why should it not also happen to other countries that are not so advanced? One reason might be that the preconditions for the demographic transition are considerably different now from what they were when the industrialized countries began their transition.

For example, prior to undergoing the demographic transition, few of the currently industrialized countries had birth rates as high as those of most currently less-developed countries, nor indeed were their levels of mortality so high. Yet when mortality did decline, it did so as a consequence of internal economic development, not as a result of a foreign country bringing in sophisticated techniques of disease prevention, as is the case today. A second reason might be that the factors leading to the demographic transition were actually different from what for years had been accepted as true. Likely it is not just change that requires explanation but also differences in the starting and ending points of the transition. Perhaps, then, the modernization theory, in and of itself, did not provide an appropriate picture of historical development. These problems with the original explanations of the demographic transition led to new research and a reformulation of the perspective.

Reformulation of the Demographic Transition Theory

One of the most important social scientific endeavors to cast doubt on the classic explanation was the European Fertility Project, directed by Ansley Coale at Princeton University. In the early 1960s, J. William Leasure, then a graduate student in economics at Princeton, was writing a doctoral dissertation on the fertility decline in Spain, using data for each of that nation's 49 provinces. Surprisingly, his thesis revealed that the history of fertility change in Spain was not explained by a simple version of the demographic transition theory. Fertility in Spain declined in contiguous areas that were culturally similar, even though the levels of urbanization and economic development might be different (Leasure 1962). At about the same time, other students began to uncover similarly puzzling historical patterns in European data (Coale 1986). A systematic review of the demographic histories of Europe was thus begun in order to establish exactly how and why the transition occurred. The focus was on the decline in fertility, because it is the most problematic aspect of the classic explanation. These new findings have been used to help revise the theory of the demographic transition.

With the discovery that the decline of fertility in Europe occurred in the context of widely differing social, economic, and demographic conditions, it became apparent that economic development may be a sufficient cause of fertility decline, but not a necessary one (Coale 1973). For example, many provinces in Europe experienced a rapid drop in their birth rate even though they were not very urban, infant mortality rates were high, and a low percentage of the population was employed in industrial occupations. The data suggest that one of the more common similarities in those areas that have undergone fertility declines is the rapid spread of **secularization.** Secularization is an attitude of autonomy from otherworldly powers and a sense of responsibility for one's own well-being (Lesthaeghe 1977; Leasure 1982; Norris and Inglehart 2004). It is associated with an enlightened view of the world—a break from traditional ways of thinking and behaving.

It is difficult to know exactly why such attitudes arise when and where they do, but we do know that industrialization and economic development are virtually always accompanied by secularization. Secularization, however, can occur independently of industrialization. It might be thought of as a modernization of thought, distinct from a modernization of social institutions. Some theorists have suggested that secularization is part of the process of westernization (see, for example, Caldwell 1982). In all events, when it pops up, secularization often spreads quickly, being diffused through social networks as people imitate the behavior of others to whom they look for clues to proper and appropriate conduct.

Education has been identified as one (indeed, probably the most important) potential stimulant to such altered attitudes, especially mass education, which tends to emphasize modernization and secular concepts. Education facilitates the rapid spread of new ideas and information, which would perhaps help explain another of the important findings from the Princeton European Fertility Project, that the onset of long-term fertility decline tended to be concentrated in a relatively short period of time (van de Walle and Knodel 1980). The data from Europe suggest that once marital fertility had dropped by as little as 10 percent in a region, the decline spread rapidly. This "tipping point" occurred whether or not infant mortality had already declined (Watkins 1986).

Some areas of Europe that were similar with respect to socioeconomic development did not experience a fertility decline at the same time, whereas other provinces that were less similar socioeconomically experienced nearly identical drops in fertility. The data suggest that this riddle is solved by examining cultural factors, not just socioeconomic ones. Building on the concept of **spatial demography** (which I will discuss in Chapter 4), it was found that areas sharing a similar culture (same language, common ethnic background, similar lifestyle) were more likely to share a decline in fertility than areas that were culturally less similar (Watkins 1991). The principal reason for this is that the idea of family planning seemed to spread quickly until it ran into a barrier to its communication. Language is one such barrier (Leasure 1962; Lesthaeghe 1977), and social and economic inequality in a region is another (Lengyel-Cook and Repetto 1982). Social distance between people turns out to inhibit communication of new ideas and attitudes.

What kinds of ideas and attitudes might encourage people to rethink how many children they ought to have? To answer this kind of question we must shift our focus from the macro (societal) level to the micro (individual) level and ask how people actually respond to the social and economic changes taking place around them. A popular individual-level perspective is that of **rational choice theory** (sometimes referred to as RAT) (Coleman and Fararo 1992). The essence of rational choice theory is that human behavior is the result of individuals making calculated cost-benefit analyses about how to act and what to do. For example, Caldwell (1976:331) has suggested that "there is no ceiling in primitive and traditional societies to the number of children who would be economically beneficial." Children are a source of income and support for parents throughout life, and they produce far more than they cost in such societies. The **wealth flow**, as Caldwell calls it, is from children to parents.

The process of modernization (the macro-level changes) eventually results in the tearing apart of large, extended family units into smaller, nuclear units that are economically and emotionally self-sufficient (micro-level changes). As that happens, children begin to cost parents more (including the cost of educating them as demanded by a modernizing society), and the amount of support that parents get from children begins to decline (starting with the income lost because children are in school rather than working). As the wealth flow reverses and parents begin to spend their income on children, rather than deriving income from them, the economic value of children vanishes. Economic rationality would now seem to dictate having zero children, but in reality, of course, people continue having children for a variety of social reasons that I detail in Chapter 6.

Rational choice theory is not just about economics, but even when it comes to economic issues, there has been a lot of research over the past few decades suggesting that we humans are not as rational as we might have thought. Daniel Kahneman, a psychologist at Princeton, won the Nobel Prize in Economics in 2002 for his work on how we think, and he summarized his research in a best-selling book, *Thinking, Fast and Slow* (Kahneman 2011). The key point is that most of our thinking is "fast" (intuitive and emotional), whereas only a small fraction is "slow" (deliberate and rational). This is what allows us to believe many things, regardless of their objective truth (Schermer 2012). It helps to explain why people do not always behave "rationally" and we have to take these ideas into account as we explain both the causes and consequences of demographic behavior. In 2002, Princeton demographer Douglas Massey suggested that people may generally be rational (with the capacity for "slow" thinking), but much of human behavior is still powered by emotional responses (the "fast" thinking) that supersede rationality (Massey 2002). We are animals, and though we may have vastly greater intellectual capacities than other species, we are still influenced by a variety of non-rational forces, including our hormones (Udry 1994, 2000).

Overall, then, the principal ingredient in the reformulation of the demographic transition perspective is to add "ideational" factors to "demand" factors as the likely causes of demographic change, especially changes in fertility. The original version of the theory suggested that modernization reduces the demand for children and so fertility falls—if people are rational economic creatures, then this is what should happen. But the real world is more complex, and the diffusion of ideas can shape fertility (and other demographic) behavior along with, or even in the absence of, the usual signs of modernization.

This does not necessarily mean that Wallerstein (1976) was correct when he declared that modernization theory was dead. On the contrary, there is evidence from around the world that "industrialization leads to occupational specialization, rising educational levels, rising income levels, and eventually brings unforeseen changes—changes in gender roles, attitudes toward authority and sexual norms; declining fertility rates; broader political participation; and less easily led publics" (Inglehart and Baker 2000:21). This is not a linear path, however. "Economic development tends to push societies in a common direction, but rather than converging, they seem to move on parallel trajectories shaped by their cultural heritages" (Inglehart and Baker 2000:49).

One strength of reformulating the demographic transition is that nearly all other perspectives can find a home here. Malthusians note with satisfaction that fertility first declined in Europe primarily as a result of a delay in marriage, much as Malthus would have preferred. Neo-Malthusians can take heart from the fact that rapid and sustained declines occurred simultaneously with the spread of knowledge about family planning practices. Marxists also find a place for themselves in the reformulated demographic transition perspective, because its basic tenet is that a change in the social structure (modernization of thought, if not also of the economy) is necessary to bring about a decline in fertility. This is only a short step away from agreeing with Marx that there is no universal law of population, but rather that each stage of development and social organization has its own law, and that cultural patterns will influence the timing and tempo of the demographic transition—when it starts and how it progresses. Furthermore, the macro-level changes are never sufficient to explain what happens—we must also pay attention to what is going on at the individual level.

The Theory of Demographic Change and Response

The work of the European Fertility Project focused on explaining regional differences in fertility declines. This was a very important theoretical development, but not a comprehensive one, because it only partially dealt with a central issue of the demographic transition theory: How (and under what conditions) can a mortality decline lead to a fertility decline? To answer that question, Kingsley Davis (1963) asked what happens to individuals when mortality declines. The answer is that more children survive through adulthood, putting greater pressure on family resources, and people have to reorganize their lives in an attempt to relieve that pressure; that is, people respond to the demographic change. But note that their response will be in terms of personal goals, not national goals. It rarely matters what a government wants. If individual members of a society do not stand to gain economically or socially by behaving in a particular way, they probably will not behave that way. Indeed, that was a major argument made by the neo-Malthusians against moral restraint. Why advocate postponement of marriage and sexual gratification rather than contraception when you know that few people who postpone marriage are actually going to postpone sexual intercourse, too? In fact, Ludwig Brentano (1910) quite forthrightly suggested that Malthus was insane to think that abstinence was the cure for the poor.

Davis argued that the response that individuals make to the population pressure created by more members joining their ranks is determined by the means available to them. A first response, nondemographic in nature, is to try to increase resources by working harder—longer hours perhaps, a second job, and so on. If that is not sufficient or there are no such opportunities, then migration of some family members (typically unmarried sons or daughters) is the easiest demographic response. This is, of course, the option that people have been using forever, undoubtedly explaining in large part why human beings have spread out over the planet.

In the early eighteenth century, Richard Cantillon, an Irish–French economist, was pointing out what happened in Europe when families grew too large (and this was even before mortality began markedly to decline):

> If all the labourers in a village breed up several sons to the same work, there will be too many labourers to cultivate the lands belonging to the village, and the surplus adults must go to seek a livelihood elsewhere, which they generally do in cities. . . . If a tailor makes all the clothes there and breeds up three sons to the same, yet there is work enough for but one successor to him, the two others must go to seek their livelihood elsewhere; if they do not find enough employment in the neighboring town they must go further afield or change their occupation to get a living. (Cantillon 1755 [1964]:23)

But what will be the response of that second generation, the children who now have survived when previously they would not have, and who have thus put the pressure on resources? Davis argues that if there is in fact a chance for social or economic improvement, then people will try to take advantage of those opportunities by avoiding the large families that caused problems for their parents. Davis suggests that the most powerful motive for family limitation is not fear of poverty or avoidance of pain as Malthus argued; rather, it is the prospect of rising prosperity that will most often motivate people to find the means to limit the number of children they have (I discuss these means in Chapter 6). Davis here echoes the themes of Mill and Dumont, but adds that, at the very least, the desire to maintain one's relative status in society may lead to an active desire to prevent too many children from draining away one's resources. Of course, that assumes the individuals in question have already attained some status worth maintaining.

One of Davis's most important contributions to our demographic perspective is, as Cicourel put it, that he "seems to rely on an implicit model of the actor who makes everyday interpretations of perceived environmental changes" (Cicourel 1974:8). For example, people will respond to a decline in mortality only if they notice it, and then their response will be determined by the social situation in which they find themselves. Davis's analysis is important in reminding us of the crucial link between the everyday lives of individuals and the kinds of population changes that take place in society. Another demographer who extended the scope of the demographic transition with this kind of analysis is Richard Easterlin.

Cohort Size Effects

People who share something in common represent a **cohort** and in population studies we usually focus especially on people who share the same age (or at least age range) in common. As I alluded to in Chapter 1, cohorts represent a potential force for change. This idea was first popularized by Norman Ryder several decades ago (Ryder 1965) with the concept of **demographic metabolism.** This refers to the ongoing replacement of people at each age in every society. Today's young people will be tomorrow's middle-aged people, and the latter will eventually replace the older population, and so on. To the extent that each cohort is different from the one that preceded it, society will change over time. Of course, people have recognized this possibility forever, which helps to explain the strict rules guiding the rearing of in most societies—tamp down innovative behavior among the young before it upsets the social order.

Societies have more trouble "tamping down" the effect of demographic metabolism when it involves a change in the size of successive cohorts, which happens as birth and or death rates (and to a lesser extent migration rates) change over time. The youth bulge, discussed in Chapter 1, is one example of that. Indeed, the impact of changing cohort size on human society is a nearly constant theme in this book. It turns out, however, that this is not just a one-way street in which, for example, changes in the birth rate alter the size of cohorts, which in turn produces social change. Richard Easterlin has shown that relative cohort size can then feed back to influence the birth rate itself.

The **Easterlin relative cohort size hypothesis** (also sometimes known as the relative income hypothesis) is based on the idea that the birth rate does not necessarily respond to absolute levels of economic well-being but rather to levels that are relative to those to which one is accustomed (Easterlin 1968, 1978). Easterlin assumes that the standard of living you experience in late childhood is the base from which you evaluate your chances as an adult. If you can easily improve your income as an adult compared to your late childhood level, then you will be more likely to marry early and have several children. If young people are relatively scarce in society and business is good, they will be in relatively high demand. In nearly classic Malthusian fashion, they will be able to command high wages and thus be more likely to feel comfortable about getting married and starting a family—the "lucky few" as Carlson (2008) has called them.

On the other hand, if young people are in relatively abundant supply, then even if business is good, the competition for jobs will be stiff and it will be difficult for people to maintain their accustomed level of living, much less marry and start a family. This is yet an another example of the youth bulge issue.

Easterlin's thesis presents a model of society in which demographic change and economic change are closely interrelated. Economic changes produce demographic changes, which in turn produce economic changes, and so on. The idea of a demographic feedback cycle, which is at the core of Easterlin's thinking, is compelling, and relative cohort size is certainly a factor that will influence various kinds of social change. But what about the situation that prevails in an increasing number of countries with relatively small cohorts of young adults who are not responding as the Easterlin hypothesis would suggest? Rather than marrying earlier and having more children, they are postponing marriage and having even fewer children. Demographers didn't see this coming, and one reaction to these unexpected trends is to suggest that parts of the world are experiencing something that goes beyond our ordinary ideas about the demographic transition.

Is There Something Beyond the Demographic Transition?

In its original formulation, the demographic transition was simply a movement from a demographic regime characterized by high birth and death rates to one characterized by low birth and death rates. When the latter was achieved, presumably the transition was over and things would stabilize demographically (homeostasis) and a country would enter a **post-transitional** era. However, the dramatic changes taking place in family and household structure since World War II, especially in Europe, led Dirk van de Kaa (1987) to talk about the "second demographic transition" as something that goes beyond a stable post-transitional period. A demographic centerpiece of this change in the richer countries (he focused on Europe) has been a fall in fertility to below-replacement levels, but van de Kaa suggested that the change was less about not having babies than it was about the personal freedom to do what one wanted, especially among women. So, rather than the pattern of grow up, marry, and have children, this transition is associated with a postponement of marriage, a rise in single living, cohabitation, and prolonged residence in the parental household (Lesthaeghe and Neels 2002; McLanahan 2004).

Chris Wilson (2013) notes that "it seems fair to conclude that the assumption of long-term convergence to replacement-level fertility has little or no basis in either empirical evidence or in demonstrably relevant theory" (p. 1375). This is in accord with van de Kaa's view that we need to revisit the notion that the end result of the "first" demographic transition is homeostasis, or population stability, replacing it with the broader view that young people increasingly make decisions about having children on the basis of self-fulfillment, without concerning themselves about biological replacement (van de Kaa 2004). We will return to this idea several times in the coming chapters, while at the same broadening the scope to include not just fertility, but also the evolving changes in mortality and migration, which are constantly churning the age structure and altering cohorts, among many other things.

The idea that change, not homeostasis, is the natural state of affairs is supported as well by the discussion in Chapter 2 about the Neolithic Demographic Transition (Bocquet-Appel 2008), which would actually represent the first demographic transition (an increase in both birth and death rates). That was followed by the more famous demographic transition of the past 200 years or so (a decrease in both birth and death rates), followed by what we would then call not the second, but rather the third demographic transition (an unexpected further decline in birth rates). We humans may have experienced not one, but a sequence of transitions, and through it all the population has generally been increasing, as shown back in Table 2.1, rather than standing still. We might more properly refer to these changes as demographic evolution, and I discuss this more thoroughly in Chapter 12 as I examine what the future might hold demographically.

The Demographic Transition Is Really a Set of Transitions

The reformulation driven by the European Fertility Project, the theory of demographic change and response, the cohort size effects, the second demographic transition, and other research all have generated the insight that the demographic transition (by which I mean the one that started a couple of hundred years ago) is actually a set of interrelated transitions. Taken together, they help us understand not just the causes but the consequences of population change. Indeed, when we view the world from this perspective, it becomes clearer why we should talk about demographic evolution, not just transition.

Usually (but not always) the first transition to occur is the **health and mortality transition** (the shift from deaths at younger ages due to communicable disease to deaths at older ages due to degenerative diseases). This transition is followed by the **fertility transition**—the shift from natural (and high) to controlled (and low) fertility, typically in a delayed response to the health and mortality transition. The predictable changes in the age structure (the **age transition**) brought about by the mortality and fertility transitions produce social and economic reactions as societies adjust to constantly changing age distributions. The rapid growth of the population occasioned by the pattern of mortality declining sooner and more rapidly than fertility almost always leads to overpopulation of rural areas, producing the **migration transition,** especially toward urban areas, creating the **urban transition.** The **family and household transition** is occasioned by the massive structural changes that accompany longer life, lower fertility, an older age structure, and urban instead of rural residence—all of which are part and parcel of the demographic transition. The interrelationships among these transitions are shown in Figure 3.4.

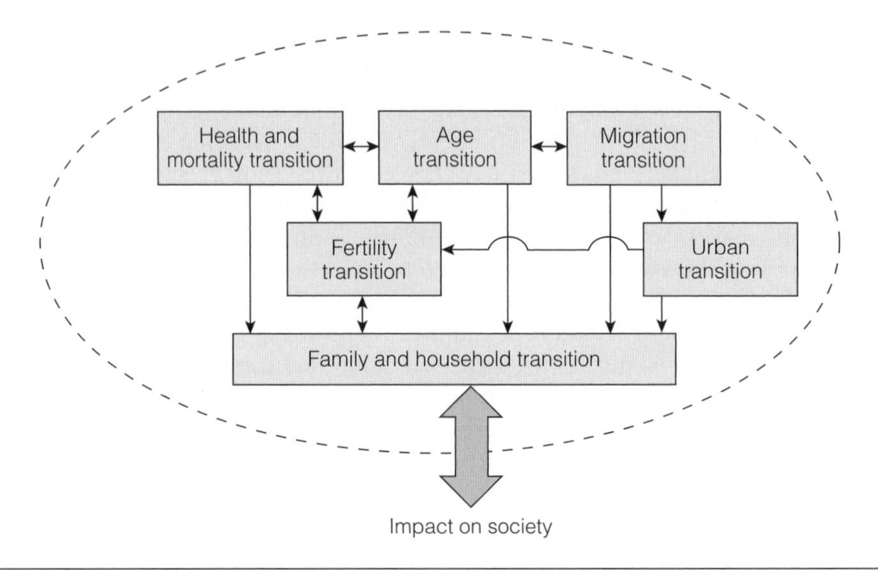

Figure 3.4 The Transitions that Comprise the Demographic Transition's Impact on Society
Note: Each box has its own set of theories that serve as explanations for the phenomenon under consideration.

The Health and Mortality Transition

The transition process almost always begins with a decline in mortality, which is brought about by changes in society that improve the health of people and thus their ability to resist disease, and by scientific advances that prevent premature death. However, death rates do not decline evenly by age; rather it is the very youngest and the very oldest—but especially the youngest—whose lives are most likely to be saved by improved life expectancy. Thus, the initial impact of the health and mortality transition is to increase the number of young people who are alive, ballooning the bottom end of the age structure in a manner that looks just like an increase in the birth rate. This sets all the other transitions in motion.

The Fertility Transition

The fertility transition can begin without a decline in mortality (as happened in France), but in most places it is the decline in mortality, leading to greater survival of children, that eventually motivates people to think about limiting the number of children they are having. Throughout most of human history, the average woman had two children who survived to adulthood. The decline in mortality, however, obviously increases that number and thereby threatens the very foundation of the household economy. At the community or societal level, the increasing number of young people creates all sorts of pressures to change, often leading to peer pressure to conform to new standards of behavior, including the deliberate control of reproduction.

Another set of extremely important changes that occur in the context of the health and mortality transition is that the scope of life expands for women as they, too, live longer. They are increasingly empowered to delay childbearing and to have fewer children because they begin to realize that most of their children will survive to adulthood and they, themselves, will survive beyond the reproductive ages, beyond their children's arrival into adulthood. This new freedom gives them vastly more opportunities than ever before in human history to do something with their lives besides bearing and raising children. This realization may be a genuine tipping point in the fertility transition, leading to an almost irreversible decline.

As fertility declines, the health and mortality transition is itself pushed along, because the survival of children is enhanced when a woman has fewer children among whom to share resources. Also affected by the fertility decline is the age structure, which begins to cave in at the younger ages as fewer children are being born, and as most are now surviving through childhood. In its turn, that shift in the age transition to an increasingly older age structure has the potential to divert societal resources away from dealing primarily with the impact of children to dealing with broader social concerns, including raising the standard of living. A higher standard of living can then redound to the benefit of health levels throughout society, adding fuel to the fire of increasing life expectancy.

The Age Transition

In many respects, the age transition is the "master" transition in that the changing number of people at each age that occurs with the decline of mortality, and then the decline in fertility, presents the most obvious demographic pressure for social change. When both mortality and fertility are high, the age structure is quite young, but the decline in mortality makes it even younger by disproportionately increasing the number of young people. Then, as fertility declines, the youngest ages are again affected first, since births occur only at age zero, so a fertility decline shows up first as simply fewer young children than before. However, as the bulge of young people born prior to the fertility decline pushes into the older ages while fertility begins to decline, the age structure moves into a stage that can be very beneficial to economic development in a society—a large fraction of the population is composed of young adults of working age who are having fewer children as dependents at the same time that the older population has not yet increased in size enough to create problems of dependency in old age. As we will see, this phase in the age transition is often associated with a golden age of advancement in the standard of living—the leap forward that Mill was advocating.

That golden age can be transitory, however, if a society has not planned for the next phase of the age transition, when the older population begins to increase more rapidly than the younger population. The baby bulge created by the initial declines in mortality reaches old age at a time when fertility has likely declined, and so the age structure has a much greater number and a higher fraction of older people than ever before. We are only now learning how societies will respond to this challenge of an increasingly older population.

The Migration Transition

Meanwhile, back at the very young age structure put into motion by declining mortality, the theory of demographic change and response suggests that in rural areas, where most of the population lived for most of human history, the growth in the number of young people will lead to an oversupply of young people looking for jobs, which will encourage people to go elsewhere in search of economic opportunity. The Europeans who experienced the first wave of population growth, because they experienced the health and mortality transition first, still lived in a world where there was "empty" land. Of course, it wasn't really empty, as the Americas and the islands of the South Pacific (largely Australia and New Zealand) were populated mainly by hunter-gatherers who used the land extensively, rather than intensively, as I mentioned in Chapter 2.

As Europeans arrived in the Americas, they perceived land as being not used and so claimed it for themselves. We all know the consequences of this for the indigenous populations, as all of the land was eventually claimed for intensive human use. So, whereas the Europeans could initially spread out from rural Europe to rural areas in the Americas and the South Pacific, migrants from rural areas today no longer have that option.

Notice in Figure 3.4 that there is a double-arrow connecting the age and migration transitions. This is because migration takes people (mainly young adults) out of one area and puts them in another area, thus affecting the age structure in both places. As I have already mentioned, this difference in age structure between two places contributes to current migration patterns of people from younger societies filling in the "empty" places in the age structure of older societies.

The Urban Transition

With empty lands filling up, migrants from the countryside in the world today have no place to go but to cities, and cities have historically tended to flourish by absorbing labor from rural areas. A majority of people in the world now live in cities, and by the end of the twenty-first century, almost all of us will be there. The urban transition thus begins with migration from rural to urban areas but then morphs into the urban "evolution" as most humans wind up being born in, living in, and dying in cities. The complexity of human existence is played out in cities, leading us to expect a constant dynamism of urban places for most of the rest of human history. Because urban places are historically associated with lower levels of fertility than rural areas, as the world's population becomes increasingly urban we can anticipate that this will be a major factor in bringing and keeping fertility levels down all over the world.

The Family and Household Transition

It is reasonable to think that the transition in family and household structure, noted above with respect to the second demographic transition, is not so much a second transition as it is another set of transitions within the broader framework of the demographic transition. As I show in Figure 3.4, the family and household transition is influenced by all the previously mentioned transitions. The health and mortality transition is pivotal because it gives women (and men, too, of course) a dramatically greater number of years to live in general, and more specifically a greater number of years that do not need to be devoted to children. Low mortality reduces the pressure for a woman to marry early and start bearing children while she is young enough for her body to handle that stress. Furthermore, when mortality was high, marriages had a high probability of ending in widowhood when one of the partners was still reasonably young, and families routinely were reconstituted as widows and widowers remarried. But low mortality leads to a much longer time that married couples will be alive together before one partner dies, and this alone is related to part of the increase in divorce rates.

The age transition plays a role at the societal level, as well, because over time the increasingly similar number of people at all ages—as opposed to a majority of people being very young—means that any society is bound to be composed of a greater array of family and household arrangements. Diversity in families and households is also encouraged by migration (which breaks up and reconstitutes families) and by the urban transition, especially since urban places tend to be more tolerant of diversity than are smaller rural communities.

The rest of this book is devoted to more detailed examinations of each of these transitions, and you will discover that a variety of theoretical approaches have developed over time to explain the causes and consequences of each set of transitions. Your own demographic perspective will be honed by looking for similarities and patterns in the transitions that link them together and, more importantly, link them to their potential impact on society.

Impact on Society

The modern field of population studies came about largely to encourage and inspire deeper insight into the causes of changes in fertility, mortality, migration, age and sex structure, and population characteristics and distribution. Demographers spent most of the twentieth century doing that, but always with an eye toward new things that could be learned about what demographic change meant for human society. Unlike in Malthus's day, population growth is no longer viewed as being caused by one set of factors nor as having a simple prescribed set of consequences.

Perhaps the closest we can come at present to "big" theories are those that try to place demographic events and behavior in the context of other global change, especially political change, economic development, and westernization. One of the more ambitious and influential of these theorists is Jack Goldstone (1991), whose work incorporates population growth as a precursor of change in the "early modern world" (defined by him as 1500 to 1800). He argues that population growth in the presence of rigid social structures produced dramatic political change in England and France, in the Ottoman Empire, and in China. Population growth led to increased government expenditures, which led to inflation, which led to fiscal crisis. In these societies with no real opportunities for social mobility, population growth (which initially increases the number of younger persons) led to disaffection and popular unrest and created a new cohort of young people receptive to new ideas. The result in the four case studies he analyzes was rebellion and revolution. This hearkens back to the discussion in Chapter 1 about the role of the "youth bulge" in conflict in the Middle East, an issue that Goldstone has examined as well (Goldstone 2002; Goldstone et al. 2011).

Stephen Sanderson has promoted the idea that population growth has been an important stimulus to change throughout human history, but especially since the Agricultural Revolution. Thus, his time frame is much longer than that of Goldstone: "Had Paleolithic hunter-gatherers been able to keep their populations from growing, the whole world would likely still be surviving entirely by hunting and gathering" (Sanderson 1995:49). Instead, population growth generated the Agricultural Revolution and then the Industrial Revolution. The sedentary life associated with the Agricultural Revolution increased social complexity (a very Durkheimian idea), which led to the rise of civilization (cities) and the state (city-states and then nation-states).

Population change may seem largely imperceptible as it is occurring, so if we can look back and see that there were momentous historical consequences of population growth and change in the past, can we look forward into the future and project similar kinds of influences? Many people (including me) would say yes, and the rest of this book will show you why.

Summary and Conclusion

A lot of thinking about population issues has taken place over a very long period, and in this chapter I have traced the progression of demographic thinking from ancient doctrines to contemporary systematic perspectives. Malthus was not the first but he was certainly the most influential of the early modern writers. Malthus believed that a biological urge to reproduce was the cause of population growth and that its natural consequence was poverty. Marx, on the other hand, did not openly argue with the Malthusian causes of growth, but he vehemently disagreed with the idea that poverty is the natural consequence of population growth. Marx denied that population growth was a problem per se—it only appeared that way in capitalist society. It may have seemed peculiar that I discuss a person who denied the importance of a demographic perspective in a chapter dedicated to that very importance. However, the Marxian point of view is sufficiently prevalent today among political leaders and intellectuals in enough countries that this attitude becomes in itself a demographic perspective of some significance. Furthermore, his perspective on the world finds its way into many aspects of current mainstream thinking, including modernization theory, that underlie aspects of the demographic transition theory.

The perspective of Mill, who seems very contemporary in many of his ideas, was somewhere between that of Malthus and Marx. He believed that increased productivity could lead to a motivation for having smaller families, especially if the influence of women was allowed to be felt and if people were educated about the possible consequences of having a large family. Dumont took these kinds of individual motivations a step further and suggested in greater detail the reasons why prosperity and ambition, operating through the principle of social capillarity, generally lead to a decline in the birth rate. Durkheim's perspective emphasized the consequences more than the causes of population growth. He was convinced that the complexity of modern societies is due almost entirely to the social responses to population growth—more people lead to higher levels of innovation and specialization.

More recently developed demographic perspectives have implicitly assumed that the consequences of population growth are serious and problematic, and they move directly to explanations of the causes of population growth. The original theory of the demographic transition suggested that growth is an intermediate stage between the more stable conditions of high birth and death rates to a new balance of low birth and death rates. Reformulations of the demographic transition perspective have emphasized its evolutionary character and have shown that the demographic transition is not one monolithic change, but rather that it encompasses several interrelated transitions: A decline in mortality will almost necessarily be followed by a decline in fertility, and by subsequent transitions in migration, urbanization, the age structure, and the family and household structure in society.

As I explore with you the causes and consequences of population growth and the uses to which such knowledge can be applied, you will need to know about the sources of demographic data. What is the empirical base of our understanding of the relationship between population and society? We turn to that topic in the next chapter.

Main Points

1. A demographic perspective is a way of relating basic population information to theories about how the world operates demographically.
2. Population doctrines and theories prior to Malthus vacillated between pronatalist and antinatalist and were often utopian.
3. According to Malthus, population growth is generated by the urge to reproduce, although growth is checked ultimately by the means of subsistence.
4. The natural consequences of population growth according to Malthus are misery and poverty because of the tendency for populations to grow faster than the food supply. Nonetheless, he believed that misery could be avoided if people practiced moral restraint—a simple formula of chastity before marriage and a delay in marriage until one can afford all the children that God might provide.
5. Marx and Engels strenuously objected to the Malthusian population perspective because it blamed poverty on the poor rather than on the evils of social organization.
6. Mill argued that the standard of living is a major determinant of fertility levels, but he also felt that people could influence their own demographic destinies.
7. Dumont argued that personal ambition generated a process of social capillarity that induced people to limit their number of children in order to get ahead socially and economically, while another French writer, Durkheim, built an entire theory of social structure on his conception of the consequences of population growth.
8. The demographic transition theory is a perspective that emphasizes the importance of economic and social development, which leads first to a decline in mortality and then, after some time lag, to a commensurate decline in fertility. It is based on the experience of the developed nations, and is derived from the modernization theory.
9. Davis's theory of demographic change and response emphasizes that people must perceive a personal need to change behavior before a decline in fertility will take place, and that the kind of response they make will depend on what means are available to them.
10. The demographic transition is really a set of transitions, including the health and mortality, fertility, age, migration, urban, and family/household transitions.

Questions for Review

1. What lessons exist within the ideas of pre-Malthusian thinkers on population that can be applied conceptually to the demographic situations we currently confront in the world?

2. It was obvious even in Malthus's lifetime that his theory had numerous defects. Describe those defects and discuss why, given them, we are still talking about Malthus.
3. Based on the information provided in this chapter, which writer—Malthus or Marx—would sound most modern and relevant to twenty-first-century demographers? Defend your answer.
4. Using the material in Chapter 2 and on the Internet as resources, reflect on the different demographic circumstances that gave rise to the Malthusian and Marxian views on population, compared to Mill, Dumont, and Durkheim. To what extent do demographic theories follow the times?
5. Review the basic premises of the theory of demographic change and response and discuss how it served to expand the concept of the demographic transition into the idea of a larger suite of transitions.

Websites of Interest

Remember that websites are not as permanent as books and journals, so I cannot guarantee that each of the following websites still exists at the moment you are reading this. You may have to Google the name of the organization to find the current web address.

1. **http://turnbull.mcs.stand.ac.uk/~history/Biographies/Condorcet.html**
   The Marquis de Condorcet, who helped to inspire Malthus's essay, is the subject of this website, located at the University of St. Andrews in Scotland. It includes biographical information and a list of his publications.
2. **http://www.efm.bris.ac.uk/het/malthus/popu.txt**
   The beauty of this website, located in the Department of Economics at the University of Bristol in England, is that it contains the full text of Malthus's first (1798) *Essay on Population.*
3. **http://www.ined.fr/en/lexicon/**
   French demographers have played key roles in developing population studies, and the National Institute of Demographic Studies (INED) in Paris carries on that tradition. At this part of the INED website, you can find a very useful glossary of demographic terms (in English), as well as data for most countries of the world.
4. **http://www.popcouncil.org**
   The Population Council is a policy-oriented research center in New York City founded in 1952 by John D. Rockefeller III. It originated the journal *Population and Development Review*, whose articles tend to focus on issues that directly or indirectly test demographic theories and perspectives. At this website you can, among other things, peruse abstracts of articles published in the journal to stay up to date on recent research.
5. **http://weekspopulation.blogspot.com/search/label/demographic%20perspectives**
   Keep track of the latest news related to this chapter by visiting my WeeksPopulation website.

CHAPTER 4
Demographic Data

Figure 4.1 Population Center of the United States Based on Data from the Decennial Censuses
Source: U.S. Census Bureau: http://www.census.gov/geo/reference/pdfs/ cenpop2010/centerpop_mean2010.pdf (accessed 2014).

SOURCES OF DEMOGRAPHIC DATA
Population Censuses
The Census of the United States
Who Is Included in the Census?
Coverage Error
Measuring Coverage Error
Content Error
Sampling Error
Continuous Measurement—American Community Survey
The Census of Canada
The Census of Mexico
IPUMS—Warehouse of Global Census Data
REGISTRATION OF VITAL EVENTS

COMBINING THE CENSUS AND VITAL STATISTICS

ADMINISTRATIVE DATA

SAMPLE SURVEYS
Demographic Surveys in the United States
Canadian Surveys
Mexican Surveys
Demographic and Health Surveys
Demographic Surveillance Systems
European Surveys

HISTORICAL SOURCES

SPATIAL DEMOGRAPHY
Mapping Demographic Data
GIS and the Census

ESSAY: Demographics of Politics: Why the Census Matters

Thus far, I have offered you a variety of facts as I described the history of population growth and provided you with an overview of the world's population situation. I do not just make up these numbers, of course, so in this chapter I discuss the various kinds of demographic data we draw on to know what is happening in the world. To analyze the demography of a particular society, we need to know how many people live there, how they are distributed geographically, how many are being born, how many are dying, how many are moving in, and how many are moving out. That, of course, is only the beginning. If we want to unravel the mysteries of why things are as they are and not just describe what they are, we have to know about the social, psychological, economic, and even physical characteristics of the people and places being studied. Furthermore, we need to know these things not just for the present but for the past as well. Let me begin the discussion, however, with sources of basic information about the numbers of living people, births, deaths, and migrants.

Sources of Demographic Data

The primary source of data on population size and distribution, as well as on demographic structure and characteristics, is the **census of population**. After an overview of the history of population censuses, I will take a closer look at how censuses are taken in the United States and its neighbors, Canada and Mexico. The major source of information on the population processes of births and deaths is the registration of **vital statistics**, although in a few countries this task is accomplished by **population registers**, and in most developing nations vital events are estimated from **sample surveys. Administrative data** and **historical data** provide much of the information about population changes at the local level and about geographic mobility and migration. Indeed, the spatial component of demography is central to our understanding of population change, and I conclude the chapter by discussing key spatial concepts.

Population Censuses

For centuries, governments have wanted to know how many people were under their rule. Rarely has their curiosity been piqued by scientific concern, but rather governments wanted to know who the taxpayers were, or they wanted to identify potential laborers and soldiers. The most direct way to find out how many people there are is to count them, and when you do that you are conducting a population census—a complete enumeration of the population. The United Nations Statistics Division (2008) notes that "the traditional census is among the most complex and massive peacetime exercises a nation undertakes. It requires mapping the entire country, mobilizing and training an army of enumerators, conducting a massive public campaign, canvassing all households, collecting individual information, compiling vast amounts of completed questionnaires, and analysing and disseminating the data" (p. 1).

In practice, this does not mean that every person is actually seen and interviewed by a census taker. In most countries, it means that one adult in a household answers questions about all the people living in that household. These answers may be verbal responses to questions asked in person by the census taker, but they also may be written responses to a questionnaire sent by mail, or even questions answered online.

The term *census* comes from the Latin for "assessing" or "taxing." For Romans, it meant a register of adult male citizens and their property for purposes of taxation, the distribution of military obligations, and the determination of political status (Starr 1987). Thus, in A.D. 119 a person named Horos from the village of Bacchias left behind a letter on papyrus in which he states: "I register myself and those of my household for the house-by-house census of the past second year of Hadrian Caesar our Lord. I am Horos, the aforesaid, a cultivator of state land, forty-eight years old, with a scar on my left eyebrow, and I register my wife Tapekusis, daughter of Horos, forty-five years old. . . ." (Winter 1936:187).

As far as we know, the earliest governments to undertake censuses of their populations were those in the ancient civilizations of Egypt, Babylonia, China, Palestine, and Rome (Bryan 2004). For several hundreds of years, citizens of Rome were counted periodically for tax and military purposes, and this enumeration of Roman subjects was extended to the entire empire, including Roman Egypt, in 5 B.C. The Bible records this event as follows: "In those days a decree went out from Caesar Augustus that all the world should be enrolled. This was the first enrollment, when Quirinius was governor of Syria. And all went to be enrolled, each to his own city" (Luke 2:1–3). You can, of course, imagine the deficiencies of a census that required people to show up at their birthplaces rather than paying census takers to go out and do the counting. And, in fact, all that was actually required was that the head of each household provide government officials with a list of every household member (Horsley 1987).

In the seventh century A.D., the Prophet Mohammed led his followers from Mecca to Medina (in Saudi Arabia), and after establishing a city-state there, one of his first activities was to conduct a written census of the entire Muslim population in the city (the returns showed a total of 1,500) (Nu'Man 1992). William of Normandy used a similar strategy in 1086, twenty years after having conquered England. William ordered an enumeration of all the landed wealth in the newly acquired territory in order to determine how much revenue the landowners owed the government. Data were recorded in the Domesday Book, *domesday* being the word in Middle English for *doomsday*, which is the day of final judgment. The census document was so named because it was the final proof of legal title to land. The Domesday Book was not really what we think of today as a census, because it was an enumeration of "hearths," or household heads and their wealth, rather than of people. In order to calculate the total population of England in 1086 from the Domesday Book, you would have to multiply the number of "hearths" by some estimate of household size. More than 300,000 households were included, and researchers estimate they averaged five persons per household. Therefore, the population the area enumerated by William at the time was approximately 1.5 million (Hinde 1998). The population of what is now modern England and Wales actually was larger than that at the time because, in fact, the Domesday Book does not cover London, Winchester, Northumberland, Durham, or much of northwest England, and the only parts of Wales included are certain border areas (U.K. National Archives 2014).

On the continent, the European renaissance began in northern Italy in the fourteenth century, and the Venetians and then the Florentines were interested in counting the wealth of their region, as William had been after conquering England. They developed a *catasto* that combined a count of the hearth and individuals. Thus, unlike the Domesday Book, the Florentine catasto of 1427 recorded not only the wealth of households but also data about each member of the household. In fact, so much information was collected that most of it went unexamined until the modern advent of computers (Herlihy and Klapisch-Zuber 1985). The value of a census was well known to François de Salignac de La Mothe-Fénelon, who was a very influential French political philosopher of the late seventeenth and early eighteenth centuries. He was the tutor to the Duke of Burgundy and much of his writing was intended as a primer of government for the young duke:

> Do you know the number of men who compose your nation? How many men, and how many women, how many farmers, how many artisans, how many lawyers, how many tradespeople, how many priests and monks, how many nobles and soldiers? What would you say of a shepherd who did not know the size of his flock? . . . A king not knowing all these things is only half a king. (quoted in Jones 2002:110)

By that description, Louis XIV (the "Sun King") and his grandson Louis XV were only partial kings, because the demographic evidence now suggests that the French population was growing in the eighteenth century, rather than declining, as the royal advisors (including the physiocrat Quesnay) believed at the time. Fénelon's books and essays were widely read in the early eighteenth century, which ushered in the modern era of nation-states, in turn giving rise to a genuine quest for accurate population information (Hollingsworth 1969). Indeed the term *statistic* is derived from the German word meaning "facts about a state." Sweden was one of the first of the European nations to keep track of its population regularly with the establishment in 1749 of a combined population register and census administered in each diocese by the local clergy (Statistika Centralbyran [Sweden] 1983). Denmark and several Italian states (before the uniting of Italy in the late nineteenth century) also conducted censuses during the eighteenth century (Carr-Saunders 1936), as did the

United States (where the first census was conducted in 1790). England launched its first modern census in 1801.

By the latter part of the nineteenth century, the statistical approach to understanding business and government affairs had started to take root in the Western world. The population census began to be viewed as a potential tool for finding out more than just how many people there were and where they lived. Governments began to ask questions about age, marital status, whether and how people were employed, literacy, and so forth. Census data (in combination with other statistics) have become the "lenses through which we form images of our society." Frederick Jackson Turner announced this famous view on the significance of the closing of the frontier on the basis of data from the 1890 census. Our national self-image today is confirmed or challenged by numbers that tell of drastic changes in the family, the increase in ethnic diversity, and many other trends. Winston Churchill observed that "first we shape our buildings and then they shape us. The same may be said of our statistics" (Alonso and Starr 1982:30).

The potential power behind the numbers that censuses produce can be gauged by public reaction to a census. In Germany, the enumeration of 1983 was postponed to 1987 because of public concern that the census was prying unduly into private lives. Germany did not conduct another census until 2002, well after reunification, and even then it was a sample census, not a complete enumeration. In the past few decades, protests have occurred in England, Switzerland, and the Netherlands, as well. In the Netherlands case, the census scheduled for the 1980s was actually canceled after a survey indicating that the majority of the urban population would not cooperate (Robey 1983). The Dutch have since used what they call a "virtual census" in which they generate population data from administrative sources, especially population registers.

In 2008, the European Union passed a set of regulations encouraging its member states to undertake census enumerations in 2011, and Germany stepped up to do that. Even before the census, German officials were concerned that they were overestimating Germany's population, and the 2011 census data confirmed that fact. Although administrative data had been capturing people moving in, out-migrants were being missed, and that became clear once the complete enumeration was undertaken.

Since the end of World War II, the United Nations has encouraged all countries to enumerate their populations in censuses, often providing financial as well as technical aid to less developed nations. The world's two largest nations, China and India, each regularly conduct censuses, with the most recent being China in 2010 and India in 2011. India is, in fact, well into its second 100 years of census taking, the first census having been taken in 1881 under the supervision of the British.

In contrast to India's regular census–taking, another of England's former colonies, Nigeria (the world's seventh most populous nation), has had more trouble with these efforts. Nigeria's population is divided among three broad ethnic groups: the Hausa-Fulani in the north, who are predominately Muslim; the Yoruba in the southwest, who are of various religious faiths; and the largely Christian Igbo in the southeast. The 1952 census of Nigeria indicated that the Hausa-Fulani had the largest share of the population, and so they dominated the first postcolonial government set up after independence in 1960. The newly independent nation ordered a census to be taken in 1962, but the results showed that northerners accounted for only 30 percent of the population. A "recount" in 1963 led somewhat suspiciously to the north accounting for 67 percent of the population. This exacerbated underlying ethnic tensions, culminating in the Igbo declaring independence. The resulting Biafran War (1967–70) saw at least 3 million people lose their lives before the Igbo rejoined the rest of Nigeria. A census in 1973 was never accepted by the government, and it was not until 1991 that the nation felt stable enough to try its hand again at enumeration, after agreeing that there would be no questions about ethnic group, language, or religion, and that population numbers would not be used as a basis for government expenditures. The official census count was 88.5 million people, well below the 110 million that many population experts had been guessing in the absence of any real data (Okolo 1999).

In March 2006, Nigeria completed its first census since 1991, but not without protests, boycotts, rows over payments to officials, and at least 15 deaths (Lalasz 2006). The final count from the 2006 census was about 140 million and, given the history of census-taking in the country, there was a lot of skepticism surrounding the numbers. The census had steered clear of questions about religion, but the 2008 Nigeria Demographic and Health Survey (see later in this chapter for a discussion of these surveys) suggests that 54 percent of women and men aged 15–49 are Christian, while about 45 percent are Muslim, and one percent practice some other religion (National Population Commission [Nigeria] and ICF Macro 2009).

Lebanon has not been enumerated since 1932, when the country was under French colonial rule (Domschke and Goyer 1986). At the time, the country's population was divided nearly equally between Christians and Muslims, and that, combined with the political strife between those groups, made taking a census a very sensitive political issue. Before the nation was literally torn apart by civil war between 1975 and 1990, the Christians had held a slight majority with respect to political representation. But Muslims almost certainly now hold a demographic majority, due both to the lower level of fertility and higher level of outmigration among Christians (Courbage and Todd 2011). Nonetheless, what we know about Lebanon comes from sources other than census data.

I should note that censuses historically have been unpopular in that part of the world. The Old Testament of the Bible tells us that in ancient times King David ordered a census of Israel in which his enumerators counted "one million, one hundred thousand men who drew the sword. . . . But God was displeased with this thing [the census], and he smote Israel. . . . So the Lord sent a pestilence upon Israel; and there fell seventy thousand men of Israel" (1 Chronicles 21). Fortunately, in modern times, the advantages of census taking seem more clearly to outweigh the disadvantages. This has been especially true in the United States, where records indicate that no census has been followed directly by a pestilence.

The Census of the United States

Population censuses were part of colonial life prior to the creation of the United States. A census had been conducted in Virginia in the early 1600s, and most of the northern colonies had conducted a census prior to the Revolution (U.S. Census Bureau 1978). As I discuss in the essay accompanying this chapter, a population census has been taken every 10 years since 1790 in the United States as part of the constitutional mandate that seats in the House of Representatives be apportioned on the basis of population size and distribution. Article 1 of the U.S. Constitution directs that "representatives. . . . shall be apportioned among the several states which may be included within this union, according to their respective numbers. . . . The actual Enumeration shall be made within three years after the first meeting of the Congress of the United States, and within every subsequent term of ten years, in such manner as they shall by law direct." Even in 1790 the government used the census to find out more than just how many people there were. The census asked for the names of the following: head of family, free white males aged 16 years and older, free white females, slaves, and other persons. The census questions were reflections of the social importance of those categories.

For the first 100 years of census taking in the United States, the population was enumerated by U.S. marshals. In 1880, special census agents were hired for the first time, and finally in 1902 the Census Bureau became a permanent part of the government bureaucracy (Hobbs and Stoops 2002). Beyond a core of inquiries designed to elicit demographic and housing information, the questions asked on the census have fluctuated according to the concerns of the time. Interest in international migration, for example, rose in 1920 just before the passage of a restrictive immigration law, and the census in that year added a battery of questions about the foreign-born population. In 2000, a question was added about grandparents as caregivers, replacing a question on fertility, and providing insight into the shift in focus from how many children women were having to the issue of who is taking care of those children. Questions are added and deleted by the Census Bureau through a process of consultation with Congress, other government officials, and census statistics users.

One of the more controversial items for the Census 2000 questionnaire was the question about race and ethnicity. The growing racial and ethnic diversity of the United States has led to a larger number of interracial/interethnic marriages and relationships producing children of mixed origin (also called "multiracial"). Previous censuses had asked people to choose a single category of race to describe themselves, but there was a considerable public sentiment that people should be able to identify themselves as being of mixed or multiple origins if, in fact, they perceived themselves in that way (Harris and Sim 2002). Late in 1997, the government accepted the recommendation from a federally appointed committee that people of mixed racial heritage be able to choose more than one race category when filling out the Census 2000 questionnaire. This was carried over into the 2010 Census, as well as incorporated into other government surveys. Thus, a person whose mother is white and whose father is African American can check both "White" and "Black or African American," whereas in the past the choice would have had to be made between the two.

There was still a separate question on "Hispanic/Latino/Spanish Origin" identity on the 2010 census, as you can see in Figure 4.2, reflecting a deep controversy that concerns the question of whether "race" is even an appropriate category to ask about. In response to a variety of concerns raised by social scientists, the Census

United States Census 2010

This is the official form for all the people at this address.
It is quick and easy, and your answers are protected by law.

U.S. DEPARTMENT OF COMMERCE
Economics and Statistics Administration
U.S. CENSUS BUREAU

Use a blue or black pen.

Start here

The Census must count every person living in the United States on April 1, 2010.

Before you answer Question 1, count the people living in this house, apartment, or mobile home using our guidelines.

- Count all people, including babies, who live and sleep here most of the time.

The Census Bureau also conducts counts in institutions and other places, so:

- Do not count anyone living away either at college or in the Armed Forces.
- Do not count anyone in a nursing home, jail, prison, detention facility, etc., on April 1, 2010.
- Leave these people off your form, even if they will return to live here after they leave college, the nursing home, the military, jail, etc. Otherwise, they may be counted twice.

The Census must also include people without a permanent place to stay, so:

- If someone who has no permanent place to stay is staying here on April 1, 2010, count that person. Otherwise, he or she may be missed in the census.

1. How many people were living or staying in this house, apartment, or mobile home on April 1, 2010?

Number of people = ☐

2. Were there any additional people staying here April 1, 2010 that you did not include in Question 1? *Mark ☒ all that apply.*
   - ☐ Children, such as newborn babies or foster children
   - ☐ Relatives, such as adult children, cousins, or in-laws
   - ☐ Nonrelatives, such as roommates or live-in baby sitters
   - ☐ People staying here temporarily
   - ☐ No additional people
3. Is this house, apartment, or mobile home — *Mark ☒ ONE box.*
   - ☐ Owned by you or someone in this household with a mortgage or loan? *Include home equity loans.*
   - ☐ Owned by you or someone in this household free and clear (without a mortgage or loan)?
   - ☐ Rented?
   - ☐ Occupied without payment of rent?
4. What is your telephone number? *We may call if we don't understand an answer.*

   Area Code + Number

OMB No. 0607-0919-C: Approval Expires 12/31/2011.

Form D-61 (1-15-2009)

5. Please provide information for each person living here. Start with a person living here who owns or rents this house, apartment, or mobile home. If the owner or renter lives somewhere else, start with any adult living here. This will be Person 1.

   What is Person 1's name? *Print name below.*

   Last Name

   First Name MI
6. What is Person 1's sex? *Mark ☒ ONE box.*
   - ☐ Male ☐ Female
7. What is Person 1's age and what is Person 1's date of birth? *Please report babies as age 0 when the child is less than 1 year old.*

   Print numbers in boxes.

   Age on April 1, 2010 Month Day Year of birth

→ NOTE: Please answer BOTH Question 8 about Hispanic origin and Question 9 about race. For this census, Hispanic origins are not races.

8. Is Person 1 of Hispanic, Latino, or Spanish origin?
   - ☐ No, not of Hispanic, Latino, or Spanish origin
   - ☐ Yes, Mexican, Mexican Am., Chicano
   - ☐ Yes, Puerto Rican
   - ☐ Yes, Cuban
   - ☐ Yes, another Hispanic, Latino, or Spanish origin — *Print origin, for example, Argentinean, Colombian, Dominican, Nicaraguan, Salvadoran, Spaniard, and so on.*
9. What is Person 1's race? *Mark ☒ one or more boxes.*
   - ☐ White
   - ☐ Black, African Am., or Negro
   - ☐ American Indian or Alaska Native — *Print name of enrolled or principal tribe.*
   - ☐ Asian Indian ☐ Japanese ☐ Native Hawaiian
   - ☐ Chinese ☐ Korean ☐ Guamanian or Chamorro
   - ☐ Filipino ☐ Vietnamese ☐ Samoan
   - ☐ Other Asian — *Print race, for example, Hmong, Laotian, Thai, Pakistani, Cambodian, and so on.* ☐ Other Pacific Islander — *Print race, for example, Fijian, Tongan, and so on.*
   - ☐ Some other race — *Print race.*
10. Does Person 1 sometimes live or stay somewhere else?
    - ☐ No ☐ Yes — *Mark ☒ all that apply.*
    - ☐ In college housing ☐ For child custody
    - ☐ In the military ☐ In jail or prison
    - ☐ At a seasonal or second residence ☐ In a nursing home
    - ☐ For another reason

→ If more people were counted in Question 1, continue with Person 2.

USCENSUSBUREAU

Figure 4.2 First Page of Questionnaire for United States Census 2010

Bureau conducted a test of the 2010 census questions using both the "standard" two questions (one on race and another on Hispanic origin) compared to a question that combined those into a single concept of "origin." The results were similar, suggesting that the single question might be a viable option (U.S. Census Bureau 2012b), although at this writing the Census Bureau has not made a decision about which version to use in the 2020 Census. Meanwhile, an article in *The Economist* (2013) nicely summarized a very complex situation:

> Such a change, say officials, would not mean that "Hispanic" is now to be considered a new racial category. Still, the widespread reporting of Hispanic-specific data, acknowledges Roberto Ramirez at the Census Bureau, means that in some respects "Hispanic" has become a de facto race.
>
> Some are sceptical about the proposal. Rubén Rumbaut, a sociologist at the University of California, Irvine, accepts the need for good data but says the bureau is thinking about

race in 18th-century terms. Hispanic identity in America, he adds, is a "Frankenstein's monster" that has taken on a life of its own.

The ethnic origins of some previous waves of immigrants have evaporated over time: Italians, Germans and Russians, dismissed by Benjamin Franklin in 1751 as of "swarthy complexion", are now, for the most part, just white. Similar forces may be at play today: last year the Pew Hispanic Centre found that among Hispanics of the third generation or above, almost half preferred to call themselves "American."

The census is designed as a complete enumeration of the population, but in the United States only a few of the questions are actually asked of everyone. For reasons of economy, most items in the census questionnaire have been administered to a sample of households in the last several censuses. From 1790 through 1930, all questions were asked of all applicable persons, but as the American population grew and Congress kept adding new questions to the census, the savings involved in sampling grew, and in 1940 the Census Bureau began its practice of asking only a small number of items of all households (the "short form"), and using a sample of one out of every six households to gather more detailed data (the "long form"). This was the procedure up through the 2000 census. A major change for the 2010 census in the United States was that it included only the short form, with the detailed data being collected not as part of the decennial census, but rather through the ongoing **American Community Survey,** which I discuss later in the chapter. Note that the short form information represents everything necessary to meet the Constitutional requirements for Congressional Redistricting. Everything else is really useful, but not Constitutionally necessary.

The questionnaire for Census 2010 is reproduced as Figure 4.2. The first page asks for a count of everyone in the household, including people who may not be there at the moment, and those who may be homeless but are nonetheless there in the housing unit. The first person listed is supposed to be someone in the household who owns, is buying, or rents this housing unit. This person used to be known as the "head of household" (and there is still a tax category in the United States for such a person), but the Census Bureau now refers to him or her on the census form simply as "Person 1." Starting then with Person 1 (which is all I show in Figure 4.2) information is requested for each person in the household regarding his or her relationship to "Person 1," sex, age and date of birth, whether the person is of Hispanic origin, and separately what the person's race is. Finally there is a specific question about whether or not this person sometimes lives or stays elsewhere. This is to help the Census Bureau eliminate duplicates, such as college students living away from home.

Table 4.1 lists the items included on the U.S. Census 2010 questionnaire (the short form) and the American Community Survey (from which "long form" data are collected), compared with a list of information obtained by the 2011 census of Canada and the 2010 census of Mexico, both of which countries are discussed later in this chapter. The table indicates which items are asked of every household and which are asked of a sample of households.

Table 4.1 Comparison of Items Included in the U.S. Census 2010 Questionnaire and the American Community Survey, the 2011 Censuses of Canada, and the 2010 Census of Mexico

Census Item	U.S. Census 2010 and ACS	Canada 2011	Mexico 2010
Population Characteristics:			
Age	XX	XX	XX
Sex	XX	XX	XX
Relationship to householder (family structure)	XX	XX	XX
Race	XX	X	
Ethnicity	XX	X	
Marital status	X	XX	XX
Fertility	X		XX
Child mortality			XX
Income	X	X	XX
Sources of income	X	X	X
Health insurance	X		XX
Job benefits			X
Unpaid household activities		X	
Labor force status	X	X	XX
Industry, occupation, and class of worker	X	X	XX
Work status last year	X		
Veteran status	X		
Grandparents as caregivers	X		
Place of work and journey to work	X	X	XX
Journey to work	X	X	
Vehicles available	X		
Ancestry	X	X	
Place of birth	X	X	XX
Birthplace of parents		X	
Citizenship	X	X	
Year of entry if not born in this country	X	X	
Language spoken at home	X	XX	XX
Language spoken at work		X	
Religion		X	XX
Educational attainment	X	X	XX
School enrollment	X	X	XX
Residence one year ago (migration)	X	X	
Residence five years ago (migration)		X	XX
International migration of family members			X
Disability (activities of daily living)	X	X	XX

Table 4.1 (continued)

Census Item	U.S. Census 2010 and ACS	Canada 2011	Mexico 2010
Housing Characteristics:			
Tenure (rent or own)	XX	XX	XX
Type of housing	XX	XX	XX
Agricultural use of property	X	XX	
Acreage of property	X		
Business use of property	X		
Material used for construction of walls			XX
Material used for construction of roof			XX
Material used for construction of floors			XX
Repairs needed on structure		X	
Year structure built	X	X	X
Units in structure	X		
Rooms in unit	X	X	XX
Bedrooms	X	X	XX
Kitchen facilities	X		XX
Electricity in house			XX
Water source			XX
Toilet facilities	X		XX
Sewerage			XX
Material possessions (TV, radio, etc.)	X		XX
House heating fuel	X		XX
Year moved into unit	X		
Value of property	X	X	
Selected housing costs	X	X	
Rent or mortgage payment	X	X	

Note: XX = Included and asked of every household; X = Included but asked of only a sample of households. Questions asked on each census may be different; similar categories of questions asked do not necessarily mean strict comparability of data.

In theory, a census obtains accurate information from everyone. But in practice that turns out to be more difficult than it may seem. For example, who is supposed to be included in the census? Are visitors to the country to be included? Are people who are absent from the country on census day to be excluded?

Who Is Included in the Census?

There are several ways to answer that question, and each produces a potentially different total number of people. At one extreme is the concept of the

de facto population, which counts people who are in a given territory on the census day. At the other extreme is the **de jure population**, which represents people who legally "belong" to a given area in some way or another, regardless of whether they were there on the day of the census. For countries with few foreign workers and where working in another area is rare, the distinction makes little difference. But many countries, including nearly all of the Gulf states in the Middle East, have large numbers of guest workers from other countries and thus have a larger de facto than de jure population. On the other hand, a country such as Mexico, from which migrants regularly leave temporarily to go to the United States, has a de jure population that is larger than the de facto population.

Most countries (including the United States, Canada, and Mexico) have now adopted a concept that lies somewhere between the extremes of de facto and de jure, and they include people in the census on the basis of **usual residence**, which is roughly defined as the place where a person usually sleeps. College students who live away from home, for example, are included at their college address rather than being counted in their parents' household. People with no usual residence (the homeless, including migratory workers, vagrants, and "street people") are counted where they are found. On the other hand, visitors and tourists from other countries who "belong" somewhere else are not included, even though they may be in the country when the census is being conducted. At the same time, the concept of usual residence means that undocumented immigrants (who legally do not "belong" where they are found) will be included in the census along with everyone else.

Where you belong became a court issue following Census 2000 in the United States. The census includes members of the military and the federal government who are stationed abroad. They are counted as belonging to the state in the United States that was their normal domicile, and in 2000 this turned out especially to benefit North Carolina, which is home to several military bases. However, Utah filed suit in federal court, objecting that Mormon missionaries from Utah who were serving abroad should also be counted as residents of Utah, rather than being excluded because they were living outside the United States. In 2001, the U.S. District Court ruled against that idea, and so North Carolina gained a seat in Congress on the basis of its "overseas residents," while Utah did not. Utah pushed the idea again for the 2010 census, but again the plan was rejected by the Census Bureau, which pointed out that an estimated 6 million Americans live abroad who are not on the U.S. government payroll and there is no reliable way to count them (Associated Press 2009).

Knowing who should be included in the census does not, however, guarantee that they will all be found and accurately counted. There are several possible errors that can creep into the enumeration process. We can divide these into the two broad categories of **nonsampling error** (which includes **coverage error** and **content error** and **sampling error**.

Coverage Error

The two most common sources of error in a census are coverage error and content error. A census is designed to count everyone, but there are always people who are

DEMOGRAPHICS OF POLITICS: WHY THE CENSUS MATTERS

Demographics are central to the political process in the United States. The constitutional basis of the Census of Population is to provide data for the **apportionment** of seats in the House of Representatives, and this process reaches down to the local level. After each enumeration, which historically takes place every ten years in the month of April, the U.S. Census Bureau is required by law to deliver total population counts for all 50 states to the president on or before December 31 of that year. These data are then sent to the House of Representatives for use in determining the number of representatives to which each state is entitled.

As the population of the United States grew and new states were added, the number of Representatives kept going up. However, since the 1910 census, the total number of House seats has been fixed at 435, and the Constitution requires that every state get at least one seat. The first 50 House seats are thus used up, taking into account the four states added subsequent to the 1910 census. The question remains of how to apportion the remaining 385 seats, remembering that congressional districts cannot cross state boundaries and there can be neither partial districts nor sharing of seats. Since 1940, the number of seats assigned to each state has been based on a formula called the method of equal proportions, which rank-orders a state's priority for each of those 385 seats based on the total population of a state compared with all other states. The calculations themselves are cumbersome, but they produce an allocation of House seats that is now accepted without much criticism or controversy, except for the issue of whether and how to count overseas Americans, as I noted elsewhere in this chapter.

The results of the 2010 census determined the number of seats for each state in the United States House of Representatives starting with the 2012 elections. This also affected the number of votes each state has in the Electoral College for the 2012 through 2020 presidential elections, since each

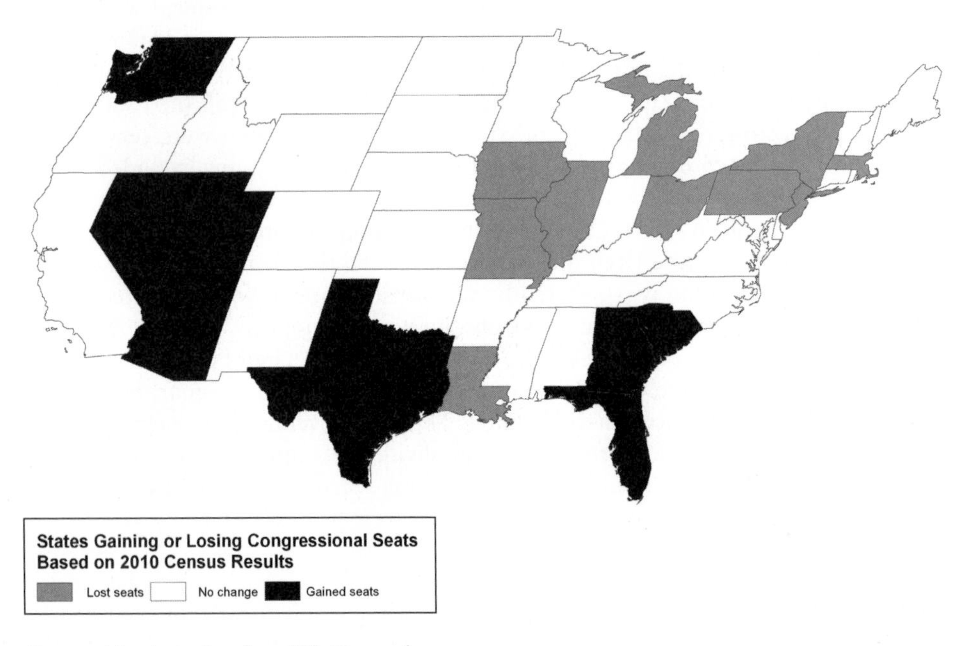

Source: Prepared by the author from U.S. Census data.

Note: The map shows only the continental states, but neither Alaska nor Hawaii lost or gained a seat as of the 2010 Census.

state's total number of electoral votes is equal to its members of Congress (Representatives and Senators combined). Because of population changes, eighteen states had changes in their number of seats following the 2010 census. Eight states gained at least one seat, and ten states lost at least one seat. The accompanying map shows where the gains and losses occurred. Texas gained the most—four seats—almost entirely on the basis on net in-migration, including many undocumented immigrants from Mexico. New York and Ohio both lost two seats, almost entirely on the basis of out-migration to southern or western states. Indeed, all but one of the states that lost at least one seat in the House of Representatives are in the northern half of the country. The lone exception is Louisiana, which lost a substantial number of people who went especially to Texas following Hurricane Katrina and have not returned. The only state for which migration was not the major contributor to population growth between 2000 and 2010 is Utah, which has the country's highest birth rate.

Once the number of seats per state has been reapportioned, the real fight begins. This involves **redistricting,** the spatial reconfiguration of congressional districts (geographic areas) that each seat will represent. The U.S. Constitution addressed the issue by requiring that " . . . The number of representatives shall not exceed one for every thirty thousand, but each state shall have at least one representative" (U.S. Constitution, Article 1, Section 2, Para 3). Thus, the framers of the Constitution were most worried about there being too few constituents per House member. This was in response to what in England at the time were known as "rotten" or "pocket" boroughs, which had plagued Parliament in England, and which were finally abolished by the Reform Act of 1832. These were boroughs in which a very small number of voters existed who could thus collude to (or be bribed to) elect a particular person to Parliament.

The framers of the Constitution decided, somewhat arbitrarily, that each Congressional District had to have a minimum of 30,000 people in order to have enough voters so that this kind of corruption would be avoided. Note that originally, those 30,000 people would have been ". . . determined by adding to the whole number of free persons, including those bound to service for a term of years, and excluding Indians not taxed, three-fifths of all other persons" (U.S. Constitution, Article 1, Section 2, Paragraph 3). The reference to "free persons" and "three-fifths of all other persons" was deleted by Section 2 of the 14th Amendment, passed in 1868 after the Civil War, and following the 13th Amendment in 1865 which abolished slavery and involuntary servitude.

In the 1960s, a series of Supreme Court decisions extended to the state and local levels the requirement that legislative districts be drawn in such a way as to ensure roughly the same number of constituents in each district ("one person, one vote"). In order to facilitate this, Public Law 94-171 mandates that the Census Bureau provide population counts by age (under 18 and 18 and over) and by race and ethnicity down to the block level for local communities (and, as noted in this chapter, these were the only data collected from everyone in the 2010 Census). These data are then used to redefine Congressional district boundaries, as well as state and local legislative boundaries.

There are very few rules that govern the formation of a district beyond the requirement that, after the census is complete, each state must change its political boundaries to make sure each congressional district is equally populated. By law, newly drawn districts do not have to take into account existing political boundaries, such as cities and counties, nor do they have to take into account natural geographic boundaries, such as mountains and rivers. They need only be equally populated. When political boundaries are drawn solely for partisan gain, the process is called "gerrymandering," after early-nineteenth-century Massachusetts governor (and later vice president of the United States under James Madison) Elbridge Gerry. Gerry attempted to draw political districts to favor his own Federalist Party over the opposing Democratic–Republicans. These districts looked like salamanders and so were dubbed "gerrymanders." Since that time, this mix of demography, geography, and politics has become a common weapon used by a

(Continued)

DEMOGRAPHICS OF POLITICS: WHY THE CENSUS MATTERS (CONTINUED)

majority party to increase its chance of winning an election by spatially clustering supportive voters together.

In 2004, the U.S. Supreme Court ruled that once the lines are drawn they can be challenged only if racial discrimination is involved. This means, in particular, that any other demographic characteristic such as political party affiliation can be used to create "safe" seats for members of Congress. If district boundaries are drawn so that one party has a clear majority within the district, then it becomes very hard for the other party to challenge the incumbent. By contrast, a "swing" district is one in which there is roughly an equal number of voters in each political party, suggesting that an incumbent could more readily be challenged. An analysis by Nate Silver following the 2012 elections in the U.S. indicated that the number of members elected from swing districts was only 35, compared to 103 members 20 years earlier. This increase in members coming from safe districts has been implicated in the increased polarization of Congress, and one of the underlying reasons for this surge in safe seats has been the uptick in gerrymandered districts (Silver 2012).

Gerrymandering is now much easier to accomplish than it used to be because GIS allows demographic data to be mapped readily, allowing an almost infinite number of possible district boundaries to be drawn and compared with one another. In most states this work is overseen by the state legislature. So, the political party in power in that state as the census data come out will make those decisions, unless the legislature has designated some other agency to do the job. Currently, non-partisan commissions undertake this task in nine states. In 2010, for example, voters in California approved Proposition 20, which added congressional redistricting to the tasks of the Redistricting Commission that already existed to redraw state legislative boundaries.

The Constitutional mandate to use the census for Congressional redistricting did not explicitly anticipate partisan politics. Over time, however, people creating boundaries of these districts have been more concerned about the demographics of voters than about the total population being served by each member of the House of Representatives. The demographic characteristics of who vote is a huge political issue partly because the definition of who can vote has shifted substantially over time. The U.S. Constitution did not set a national standard for voter eligibility, leaving that decision to the states. Most states adopted the principle that only white male landowners were eligible to vote, and it was not until the middle of the nineteenth century that land ownership requirements were removed in all states, expanding the electorate to *all* white males.

The assumption is that you must be a citizen to vote and all states do require that, although it is not explicitly stated in the Constitution. This became an issue after the Civil War when southern states wanted to exclude former slaves from being able to vote by claiming that they were not U.S. citizens. So, just who ***is*** a citizen? That issue was settled by the 14th Amendment to the Constitution in 1868, which in Section 1 says that: "All persons born or naturalized in the United States and subject to the jurisdiction thereof, are citizens of the United States." This was followed in 1870 by the 15th Amendment, which provides that a person is eligible to vote regardless of race (and this was further reinforced by the Voting Rights Act of 1965). It was not until 1920 that women were given the vote by the passage of the 19th Amendment, and not until 1924 that all American Indians were granted the right to vote. Most recently, in 1971, the 26th Amendment lowered the voting age throughout the U.S. from 21 to 18. You can see, then, that over time there has been a dramatic convergence between the number of people living in a Congressional District (which determines how the boundaries of a District are defined) and the number of potential voters in that District (the people who will actually elect the Representative from that District).

Discussion Questions: (1) Discuss how the purpose for which a census is taken may influence the kinds of questions asked, and the methodology used to collect those data; **(2)** What are some ways in which the redistricting procedures called "gerrymandering" could be limited? What might be the unintended consequences of your suggested changes?

missed, as well as some who are counted more than once. The combination of the undercount and the overcount is called coverage error, or **net census undercount** (the difference between the undercount and the overcount). There are several ways to measure and adjust for undercount, but it becomes more complicated (and political) when there is a **differential undercount**, in which some groups are more likely to be underenumerated than others.

In the United States, the differential undercount has meant that racial/ethnic minority groups (especially African Americans) have been less likely to be included in the census count than whites. Table 4.2 shows estimates of the net undercount in the last several censuses, along with the differential undercount of the black population. The overall undercount in the 1940 census was 5.6 percent, and you can see that it has been steadily declining since then as the Census Bureau institutes ever more sophisticated procedures. But you can also see that in 1940 more than 10 percent of African Americans in the country were missed by the census. This was the year that the differential undercount was discovered as a result of a "somewhat serendipitous natural experiment" (Anderson and Fienberg 1999:29). Because of World War II, men were registering for the draft when that census was taken, providing demographers with a chance to compare census returns with counts of men registering for the draft. It turned out that 229,000 more black men signed up for the draft than would have been expected based on census data (Price 1947), signaling some real problems with the completeness of the census coverage in 1940. Since then, a great deal of time, effort, and controversy have gone into attempts to reduce both the overall undercount and the differential undercount. The numbers

Table 4.2 Net Undercount and Differential Undercount in U.S. Censuses

Year	Net undercount for total population (%)	Undercount of black population (%)	Undercount of white population (%)	Differential undercount (percentage point difference between black and white undercount)
1940	5.6	10.3	5.1	5.2
1950	4.4	9.6	3.8	5.8
1960	3.3	8.3	2.7	5.6
1970	2.9	8.0	2.2	5.8
1980	1.4	5.9	0.7	5.2
1990	1.8	5.7	1.3	4.4
2000	(0.5)	1.8	(1.1)	2.9
2010	0.0	2.1	(0.8)	2.9

Sources: Data for 1940 through 1980 are from Anderson and Fienberg (1999: Table 4.1); data for 1990 are from Robinson, West, and Adlakha (2002: Table 6), and data for 2000 and 2010 are from the U.S. Census Bureau (2012a). The undercounts for 1940 through 1990 are based on demographic analysis, and the undercounts for 2000 and 2010 are based on Post-Enumeration Surveys.

Note: Numbers in parentheses indicate a net overcount; for 2000 and 2010 the white population refers to the non-Hispanic white only population.

in Table 4.2 show that Census 2010 appears to have been more successful than any previous census in this regard, with the undercount of some groups being balanced by overcount in others. Thus, blacks were still slightly undercounted while non-Hispanic whites were slightly overcounted.

Coverage is improved in the census by a variety of measures, such as having better address identification so that every household receives a questionnaire and having a high-profile advertising campaign designed to encourage a high response to the mail-out questionnaire. Nearly three-fourths (72 percent) of households responded to the mailed questionnaire in 2010, and the rest were contacted by the Census Bureau in the Non-Response Follow-Up (NRFU) phase of data collection. The Census Bureau sends staff members into the field to interview people who do not complete the forms, and, in some cases, to find out about people whom they were unable to contact. When you combine this with the fact that one member of a household may have filled in the information for all household members, it is easy to see why so many people routinely think they have not been counted in the census—someone else answered for them.

In China, coverage error has focused not on racial/ethnic groups but on children. Goodkind (2011) estimates that there were nearly 37 million children under the age of ten who were not counted in the 2000 census in China. The reason for this was not that the census takers could not find them, but rather that they were being hidden. Acknowledging them would have provided evidence that the government-mandated birth quotas had been exceeded. Since local officials, not just parents, were held responsible for failure to keep the birth rate down, everybody at the local level had an interest in suppressing information about these children. A related issue with coverage error is that it is dependent upon the definition of who should be counted. In 2000, the Chinese census enumerated only those people with Chinese citizenship, whereas in 2010 the Chinese shifted to counting people who usually reside in China—thus including immigrants (Feng 2012).

Measuring Coverage Error

Right now you are probably asking yourself how a country's Census Bureau could ever begin to estimate the number of people missed in a census. This is not an easy task, and statisticians in the United States and other countries have experimented with a number of methods over the years. The two principal methods used are (1) **demographic analysis (DA)**, and (2) **dual-system estimation (DSE)** (which typically involves a post-enumeration survey).

The demographic analysis approach uses the **demographic balancing equation** to estimate what the population at the latest census should have been, and then compares that number to the actual count. The demographic balancing equation says that the population at time 2 is equal to the population at time 1 plus the births between time 1 and 2, minus the deaths between time 1 and 2, plus the in-migrants between time 1 and 2, minus the out-migrants between time 1 and 2. Thus, if we know the number of people from the previous census, we can add the number of births since then, subtract the number of deaths since then, add the number of in-migrants since then, and subtract the number of out-migrants since then to estimate what the total population count should have been. A comparison of this number with the actual census count provides a clue as to the accuracy of the census. Using these methods, the Census Bureau is able to piece together a composite rendering of what the population "should" look like. Differences from that picture and the one painted by the census can be used as estimates of under- or over-enumeration. By making these calculations for all age, sex, and racial/ethnic groups, we can arrive at an estimate of the possible undercount among various groups in the population. This, for example, was the basis for deciding that China's 2000 census had missed 37 million children.

Of course, if we do not have an accurate count of births, deaths, and migrants, then our demographic-analysis estimate may itself be wrong, so this method requires careful attention to the quality of the non-census data. And, you say, why should we even take a census if we think we can estimate the number of people more accurately without it? The answer is that the demographic-analysis approach usually only produces an estimate of the total number of people in any age, sex, racial/ethnic group, without providing a way of knowing the details of the population—which is what we obtain from the census questionnaire.

The dual-system estimation method involves comparing the census results with some other source of information about the people counted. For example, after Census 2010 in the United States, the Census Bureau implemented its Accuracy and Coverage Evaluation (A.C.E.) Survey, which was similar to, albeit somewhat more complex than, the post-enumeration A.C.E survey conducted after the 2000 census, because it incorporated a variety of innovations suggested by a committee of the National Research Council (Bell and Cohen 2008). The process involves taking a carefully constructed sample survey right after the census is finished and then matching people in the sample survey with their responses in the census. This process can determine whether households and individuals within the households were counted in both the census and the survey (the ideal situation); in the census but not in the survey (possible but not likely); or in the survey but not in the census (the usual measure of underenumeration), or counted in the wrong place. Obviously, some people may be missed by both the census and the survey, but the logic underlying the method is analogous to the capture-recapture method used by biologists tracking wildlife (Choldin 1994). That strategy is to capture a sample of animals, mark them, and release them. Later, another sample is captured, and some of the marked animals will wind up being recaptured. The ratio of recaptured animals to all animals caught in the second sample is assumed to represent the ratio of the first group captured to the whole population, and on this basis the wildlife population can be estimated. Although some humans are certainly "wild," a few adjustments are required to apply the method to human populations.

Content Error

Coverage error is a concern in any census, but there can also be problems with the accuracy of the data obtained in the census (content error). Content error includes nonresponses to particular questions on the census or inaccurate responses if people do not understand the question. Errors can also occur if information is inaccurately recorded on the form or if there is some glitch in the processing (coding, data entry, or editing) of the census return. By and large, content error seems not to be a problem in the U.S. census, although the data are certainly not 100 percent accurate. There is always the potential for misunderstanding the meaning of a question, and these problems appear to be greater for people with lower literacy skills (Iversen et al. 1999). In general, data from the United Nations suggest that the more highly developed a country is, the more accurate the content of its census data will be, and this is probably accounted for largely by higher levels of education.

In less developed countries, content error may be more problematic because interviewers may not be sufficiently trained or motivated to press respondents for accurate information. Over the years, a seemingly simple question such as age has been prone to error because of "age-heaping" in which people round their age to the nearest zero or five instead of giving a precise answer. This is why the U.S. Census asks about birth date (see Figure 4.2), not simply your age.

Sampling Error

If any of the data in a census are collected on a sample basis (as is done, for example, in the United States, Canada, and Mexico), then sampling error is introduced into the results. With any sample, scientifically selected or not, differences are likely to exist between the characteristics of the sampled population and the larger group from which the sample was chosen. However, in a scientific sample, such as that used in most census operations, sampling error is readily measured based on the mathematics of probability. To a certain extent, sampling error can be controlled—samples can be designed to ensure comparable levels of error across groups or across geographic areas. Furthermore, if the sample is very large, then sampling error will be relatively small. Nonsampling error and the biases it introduces throughout the census process probably reduce the quality of results more than sampling error (Schneider 2003).

Continuous Measurement—American Community Survey

Almost all the detailed data about population characteristics obtained from the decennial censuses in the United States come from the "long form," which for several decades was administered to about one in six households. The success of survey sampling in obtaining reliable demographic data led the U.S. Census Bureau in 1996 to initiate a process of "continuous measurement" designed to replace the long form in subsequent decennial censuses, beginning with the 2010 census (Torrieri 2007). The vehicle for this is the monthly American Community Survey (ACS), which is a "rolling survey" of approximately 3 million American households each year, designed to collect enough data over a ten-year period to provide detailed information down to the census block level, and in the process provide updated information on an annual basis, rather than having to wait for data at ten-year intervals. Just as with the census, questionnaires are mailed out to the households selected for the sample, and if they are not returned, the data are collected by phone, or by a personal visit from the Census Bureau (Griffin and Waite 2006).

The first data from the American Community Survey were made available on the Internet for the 2005 round of data, and data for the nation, states, and large populations within states are now updated annually online at http://factfinder2.census.gov. By 2010 enough surveys had been collected to produce five-year estimates for areas with populations less than 20,000. There are about 115 million households in the United States, and 3 million of those are surveyed each year, so over the course of its ten years between the 2000 and 2010 censuses, the ACS encompassed about 30 million households, representing slightly more than one in four households that wound up providing the detailed data no longer collected in the decennial census. The five-year estimates from the ACS produce data similar to that generated by Canada's five-year census cycle.

The Census of Canada

The first census in Canada was taken in 1666 when the French colony of New France was counted on the order of King Louis XIV (perhaps under Fénelon's influence, as I mentioned earlier). This turned out to be a door-to-door enumeration of all 3,215 settlers in Canada at that time. A series of wars between England and France ended with France ceding Canada to England in 1763, and the British undertook censuses on an irregular but fairly consistent basis. The several regions of Canada were united under the British North America Act of 1867, which specified that censuses were to be taken regularly to establish the number of representatives that each province would send to the House of Commons. The first of these was taken in 1871, although similar censuses had been taken in 1851 and 1861. In 1905, the census bureau became a permanent government agency, now known as Statistics Canada.

Canada began using sampling in 1941, the year after the United States experimented with it. In 1956, Canada conducted its first quinquennial census (every five years, as opposed to every ten years—the decennial census), and in 1971 Canada mandated that the census be conducted every five years. The U.S. Congress passed similar legislation in the 1970s but never funded those efforts, so the United States stayed with the decennial census until the implementation of continuous measurement provided by the American Community Survey.

As in the U.S., two census forms had been used in Canada from 1941 through the 2006 census—a short form for all households with just a few key items (see Table 4.1) and a more detailed long form that went to a sample of 20 percent of Canadian households. However, for the 2011 Census, Statistics Canada dropped the mandatory long form and instead implemented the National Household Survey. This was sent out to 30 percent of households with the caveat that responding was voluntary, rather than compulsory. The Director of Statistics Canada resigned after the government made that decision, and there is concern that the high nonresponse rate (26 percent nationally but higher in some provinces) may make these data very difficult to interpret (Hulchanski et al. 2013).

Canada's population is even more diverse than that of the United States and so the mandatory short form asks several questions about language—indeed, the split between English and French speakers nearly tore the country apart in the 1990s. The National Household Survey includes detailed questions about race/ethnicity, place of birth, citizenship, and ancestry.

Statistics Canada estimates coverage error by comparing census results with population estimates (the demographic analysis approach) and by conducting a Reverse Record Check study to measure the undercoverage errors and also an Overcoverage Study designed to investigate overcoverage errors. The Reverse Record Check is the most important part of this, and involves taking a sample of records from other sources such as birth records and immigration records and then looking for those people in the census returns. An analysis of people not found who should have been there is a key component of estimating coverage error. The results of the Reverse Record Check and the Overcoverage Study are then combined to provide an estimate of net undercoverage, which was 2.3 percent in the 2011 census, compared to 2.8 percent in 2006 (Statistics Canada 2013a).

The Census of Mexico

Like Canada and the United States, Mexico has a long history of census taking. There are records of a census in the Valley of Mexico taken in the year 1116, and the subsequent Aztec empire also kept count of the population for tax purposes. Spain conducted several censuses in Mexico during the colonial years, including a general census of New Spain (Nueva España, as they knew it) in 1790. Mexico gained independence from Spain in 1821, but it was not until 1895 that the first of the modern series of national censuses was undertaken. A second enumeration was done in 1900, and since then censuses have been taken every 10 years (with the exception of the one in 1921, which was one year out of sequence because of the Mexican civil war—*aka* Mexican Revolution—between 1910 and 1920). From 1895 through the 1970s, the census activities were carried out by the General Directorate of Statistics (Dirección General de Estadística), and there were no permanent census employees. However, the bureaucracy was reorganized for the 1980 census, and in 1983 the Instituto Nacional de Estadística, Geografía e Informática (INEGI) became the permanent government agency in charge of the census and other government data collection.

A somewhat different set of questions is asked in Mexican censuses than in those of the United States and Canada, as you can see in Table 4.1. The 2000 census was the first in Mexico to use a combination of a basic questionnaire administered to most households, plus a lengthier questionnaire administered to a sample of households, and this was replicated in the 2010 census. Furthermore, the sampling strategy was a bit different than in the United States and Canada. Most of the questions are asked of all households, and the sample involves asking 10 percent of households to respond to a set of more detailed questions about topics included in the basic questionnaire. Especially noteworthy is a set of questions seeking information about family members who had been international migrants at any time during the previous five years. In 1995 and again in 2005, Mexico conducted a mid-decade census, which it calls a "Conteo," to distinguish it from the decennial censuses. The Conteo uses only the basic questionnaire, and does not include a sample to receive the extended questionnaire. In Mexico, all census forms are administered in person by census takers hired by INEGI. Neither mail-back nor Internet forms are yet available.

Less income detail is obtained in Mexico than in Canada or the United States, and socioeconomic categories are more often derived from outward manifestations of income, such as housing quality, and material possessions owned by members of the household, about which there are several detailed questions. Because most Mexicans are *mestizos* (Spanish for mixed race, in this case mainly European and indigenous), no questions are asked about race or ethnicity. The only allusion to diversity within Mexico on the basic questionnaire is found in the question about language, in which people are asked if they speak an Indian language. If so, they are also asked if they speak Spanish. On the long form administered to a sample of households, a question is also asked specifically about whether or not they belong to an indigenous group.

In Mexico, the evaluation of coverage error in the census has generally been made using the method of demographic analysis. On this basis, Corona Vásquez (1991) estimated that underenumeration in the 1990 Mexican census was somewhere between 2.3 and 7.3 percent. No analysis has been published of the accuracy of subsequent censuses, which, in all events, would be difficult to establish because of the large number of Mexican nationals living outside of the country, especially in the United States. Perhaps more important is the fact that post-enumeration surveys and other types of coverage error analyses are expensive and, as it was, budget cuts forced INEGI to trim several questions from the extended questionnaire administered in the 2010 census.

IPUMS—Warehouse of Global Census Data

It is taken for granted in North America that census data can be downloaded for free from the Internet. The United States, Canada, and Mexico all provide such access, as do an increasing number of countries around the world. In all cases, however, the data are already tabulated for you by the statistical agencies. For researchers interested in uncovering trends and patterns in the data, it is vastly preferable to have access to data from the individual census records so that detailed statistical analysis can be undertaken, as long you have the requisite statistical software such as SPSS, SAS, or STATA. Census agencies provide these kinds of data by creating what are known as **Public Use Microdata Samples (PUMS)**. A small sample of all census records, typically 5 to 10 percent, is randomly selected. These records are stripped of all personally identifying information, but with a geographic code left in place so that a person's general location can be determined, and then this data set is made available to researchers for analysis.

Over the past two decades, the Minnesota Population Center at the University of Minnesota has been creating a genuinely amazing resource of public use microdata samples from the censuses of the United States (1850 to the present), the American Community Survey from 2001 to the present, the Current Population Survey of the United States (discussed later in the chapter) from 1962 to the present, and an ever-growing library of data from censuses all over the globe. The files are harmonized so that variable definitions are similar from one census to another, and they are provided in standard statistical software formats, along with links to digital maps for those countries. The Minnesota Population Center also hosts the National Historical Geographic Information System which includes georeferenced aggregated (not individual level) U.S. census data from 1790 to the present, all linked to digital maps that you can download and analyze yourself. They have several other related projects, and overall these are the kinds of data that truly move us forward in our understanding of what's going on in the world. I encourage you to go to www.ipums.org to investigate this resource.

Registration of Vital Events

If you were born in the United States, a birth certificate was filled out for you, probably by a clerk or volunteer staff person in the hospital where you were born. When you die, someone (again, typically a hospital clerk) will fill out a death certificate on your behalf. Standard birth and death certificates used in the United States are shown in Figure 4.3. Births and deaths, as well as marriages, divorces, and abortions, are known as vital events, and when they are recorded by the government and compiled for use they become vital statistics. These statistics are the major source of data on births and deaths in most countries, and they are most useful when combined with census data.

Registration of vital events in Europe actually began as a chore of the church. Priests often recorded baptisms, marriages, and deaths, and historical demographers have used the surviving records to reconstruct the demographic history of parts of Europe (Wrigley 1974; Wrigley and Schofield 1981; Wall et al. 1983; Landers 1993). Among the more demographically important tasks that befell the clergy was that of recording burials that occurred in England during the many years of the plague. In the early sixteenth century, the city of London ordered that the number of people dying be recorded in each parish, along with the number of christenings. Beginning in 1592, these records (or "bills") were printed and circulated on a weekly basis during particularly rough years, and so they were called the London Bills of Mortality (Laxton 1987). Between 1603 and 1849, these records were published weekly (on Thursdays, with an annual summary on the Thursday before Christmas) in what amounts to one of the most important sets of vital statistics prior to the nineteenth-century establishment of official government bureaucracies to collect and analyze such data.

U.S. STANDARD CERTIFICATE OF LIVE BIRTH

LOCAL FILE NO. BIRTH NUMBER:

CHILD
1. CHILD S NAME (First, Middle, Last, Suffix) | 2. TIME OF BIRTH (24hr) | 3. SEX | 4. DATE OF BIRTH (Mo/Day/Yr)
5. FACILITY NAME (If not institution, give street and number) | 6. CITY, TOWN, OR LOCATION OF BIRTH | 7. COUNTY OF BIRTH

MOTHER
8a. MOTHER S CURRENT LEGAL NAME (First, Middle, Last, Suffix) | 8b. DATE OF BIRTH (Mo/Day/Yr)
8c. MOTHER S NAME PRIOR TO FIRST MARRIAGE (First, Middle, Last, Suffix) | 8d. BIRTHPLACE (State, Territory, or Foreign Country)
9a. RESIDENCE OF MOTHER-STATE | 9b. COUNTY | 9c. CITY, TOWN, OR LOCATION
9d. STREET AND NUMBER | 9e. APT. NO. | 9f. ZIP CODE | 9g. INSIDE CITY LIMITS? □ Yes □ No

FATHER
10a. FATHER S CURRENT LEGAL NAME (First, Middle, Last, Suffix) | 10b. DATE OF BIRTH (Mo/Day/Yr) | 10c. BIRTHPLACE (State, Territory, or Foreign Country)

CERTIFIER
11. CERTIFIER S NAME: ______
TITLE: □ MD □ DO □ HOSPITAL ADMIN. □ CNM/CM □ OTHER MIDWIFE □ OTHER (Specify)______
12. DATE CERTIFIED __/__/____ MM DD YYYY | 13. DATE FILED BY REGISTRAR __/__/____ MM DD YYYY

INFORMATION FOR ADMINISTRATIVE USE

MOTHER
14. MOTHER S MAILING ADDRESS: □ Same as residence, or: State: City, Town, or Location:
Street & Number: Apartment No.: Zip Code:
15. MOTHER MARRIED? (At birth, conception, or any time between) □ Yes □ No
IF NO, HAS PATERNITY ACKNOWLEDGMENT BEEN SIGNED IN THE HOSPITAL? □ Yes □ No
16. SOCIAL SECURITY NUMBER REQUESTED FOR CHILD? □ Yes □ No | 17. FACILITY ID. (NPI)
18. MOTHER S SOCIAL SECURITY NUMBER: | 19. FATHER S SOCIAL SECURITY NUMBER:

INFORMATION FOR MEDICAL AND HEALTH PURPOSES ONLY

MOTHER

20. MOTHER S EDUCATION (Check the box that best describes the highest degree or level of school completed at the time of delivery)
- □ 8th grade or less
- □ 9th - 12th grade, no diploma
- □ High school graduate or GED completed
- □ Some college credit but no degree
- □ Associate degree (e.g., AA, AS)
- □ Bachelor s degree (e.g., BA, AB, BS)
- □ Master s degree (e.g., MA, MS, MEng, MEd, MSW, MBA)
- □ Doctorate (e.g., PhD, EdD) or Professional degree (e.g., MD, DDS, DVM, LLB, JD)

21. MOTHER OF HISPANIC ORIGIN? (Check the box that best describes whether the mother is Spanish/Hispanic/Latina. Check the No box if mother is not Spanish/Hispanic/Latina)
- □ No, not Spanish/Hispanic/Latina
- □ Yes, Mexican, Mexican American, Chicana
- □ Yes, Puerto Rican
- □ Yes, Cuban
- □ Yes, other Spanish/Hispanic/Latina (Specify)______

22. MOTHER S RACE (Check one or more races to indicate what the mother considers herself to be)
- □ White
- □ Black or African American
- □ American Indian or Alaska Native (Name of the enrolled or principal tribe)______
- □ Asian Indian
- □ Chinese
- □ Filipino
- □ Japanese
- □ Korean
- □ Vietnamese
- □ Other Asian (Specify)______
- □ Native Hawaiian
- □ Guamanian or Chamorro
- □ Samoan
- □ Other Pacific Islander (Specify)______
- □ Other (Specify)______

FATHER

23. FATHER S EDUCATION (Check the box that best describes the highest degree or level of school completed at the time of delivery)
- □ 8th grade or less
- □ 9th - 12th grade, no diploma
- □ High school graduate or GED completed
- □ Some college credit but no degree
- □ Associate degree (e.g., AA, AS)
- □ Bachelor s degree (e.g., BA, AB, BS)
- □ Master s degree (e.g., MA, MS, MEng, MEd, MSW, MBA)
- □ Doctorate (e.g., PhD, EdD) or Professional degree (e.g., MD, DDS, DVM, LLB, JD)

24. FATHER OF HISPANIC ORIGIN? (Check the box that best describes whether the father is Spanish/Hispanic/Latino. Check the No box if father is not Spanish/Hispanic/Latino)
- □ No, not Spanish/Hispanic/Latino
- □ Yes, Mexican, Mexican American, Chicano
- □ Yes, Puerto Rican
- □ Yes, Cuban
- □ Yes, other Spanish/Hispanic/Latino (Specify)______

25. FATHER S RACE (Check one or more races to indicate what the father considers himself to be)
- □ White
- □ Black or African American
- □ American Indian or Alaska Native (Name of the enrolled or principal tribe)______
- □ Asian Indian
- □ Chinese
- □ Filipino
- □ Japanese
- □ Korean
- □ Vietnamese
- □ Other Asian (Specify)______
- □ Native Hawaiian
- □ Guamanian or Chamorro
- □ Samoan
- □ Other Pacific Islander (Specify)______
- □ Other (Specify)______

Mother s Name______ Mother s Medical Record No.______

26. PLACE WHERE BIRTH OCCURRED (Check one)
- □ Hospital
- □ Freestanding birthing center
- □ Home Birth: Planned to deliver at home? □ Yes □ No
- □ Clinic/Doctor s office
- □ Other (Specify)______

27. ATTENDANT S NAME, TITLE, AND NPI
NAME: ______ NPI:____
TITLE: □ MD □ DO □ CNM/CM □ OTHER MIDWIFE □ OTHER (Specify)______

28. MOTHER TRANSFERRED FOR MATERNAL MEDICAL OR FETAL INDICATIONS FOR DELIVERY? □ Yes □ No
IF YES, ENTER NAME OF FACILITY MOTHER TRANSFERRED FROM: ______

REV. 11/2003

MOTHER
29a. DATE OF FIRST PRENATAL CARE VISIT __/__/____ MM DD YYYY □ No Prenatal Care | 29b. DATE OF LAST PRENATAL CARE VISIT __/__/____ MM DD YYYY | 30. TOTAL NUMBER OF PRENATAL VISITS FOR THIS PREGNANCY ______ (If none, enter "0".)
31. MOTHER S HEIGHT ____ (feet/inches) | 32. MOTHER S PREPREGNANCY WEIGHT ____ (pounds) | 33. MOTHER S WEIGHT AT DELIVERY ____ (pounds) | 34. DID MOTHER GET WIC FOOD FOR HERSELF DURING THIS PREGNANCY? □ Yes □ No

35. NUMBER OF PREVIOUS LIVE BIRTHS (Do not include this child)
- 35a. Now Living: Number ____ □ None
- 35b. Now Dead: Number ____ □ None

36. NUMBER OF OTHER PREGNANCY OUTCOMES (spontaneous or induced losses or ectopic pregnancies)
- 36a. Other Outcomes: Number ____ □ None

37. CIGARETTE SMOKING BEFORE AND DURING PREGNANCY
For each time period, enter either the number of cigarettes or the number of packs of cigarettes smoked. IF NONE, ENTER "0".
Average number of cigarettes or packs of cigarettes smoked per day.

	# of cigarettes		# of packs
Three Months Before Pregnancy	____	OR	____
First Three Months of Pregnancy	____	OR	____
Second Three Months of Pregnancy	____	OR	____
Third Trimester of Pregnancy	____	OR	____

38. PRINCIPAL SOURCE OF PAYMENT FOR THIS DELIVERY
- □ Private Insurance
- □ Medicaid
- □ Self-pay
- □ Other (Specify)______

35c. DATE OF LAST LIVE BIRTH __/____ MM YYYY | 36b. DATE OF LAST OTHER PREGNANCY OUTCOME __/____ MM YYYY | 39. DATE LAST NORMAL MENSES BEGAN __/__/____ MM DD YYYY | 40. MOTHER S MEDICAL RECORD NUMBER

MEDICAL AND HEALTH INFORMATION

41. RISK FACTORS IN THIS PREGNANCY (Check all that apply)
- Diabetes
  - □ Prepregnancy (Diagnosis prior to this pregnancy)
  - □ Gestational (Diagnosis in this pregnancy)
- Hypertension
  - □ Prepregnancy (Chronic)
  - □ Gestational (PIH, preeclampsia)
  - □ Eclampsia
- □ Previous preterm birth
- □ Other previous poor pregnancy outcome (Includes perinatal death, small-for-gestational age/intrauterine growth restricted birth)
- □ Pregnancy resulted from infertility treatment-If yes, check all that apply:
  - □ Fertility-enhancing drugs, Artificial insemination or Intrauterine insemination
  - □ Assisted reproductive technology (e.g., in vitro fertilization (IVF), gamete intrafallopian transfer (GIFT))
- □ Mother had a previous cesarean delivery If yes, how many ______
- □ None of the above

42. INFECTIONS PRESENT AND/OR TREATED DURING THIS PREGNANCY (Check all that apply)
- □ Gonorrhea
- □ Syphilis
- □ Chlamydia
- □ Hepatitis B
- □ Hepatitis C
- □ None of the above

43. OBSTETRIC PROCEDURES (Check all that apply)
- □ Cervical cerclage
- □ Tocolysis
- External cephalic version:
  - □ Successful
  - □ Failed
- □ None of the above

44. ONSET OF LABOR (Check all that apply)
- □ Premature Rupture of the Membranes (prolonged, 12 hrs.)
- □ Precipitous Labor (<3 hrs.)
- □ Prolonged Labor (20 hrs.)
- □ None of the above

45. CHARACTERISTICS OF LABOR AND DELIVERY (Check all that apply)
- □ Induction of labor
- □ Augmentation of labor
- □ Non-vertex presentation
- □ Steroids (glucocorticoids) for fetal lung maturation received by the mother prior to delivery
- □ Antibiotics received by the mother during labor
- □ Clinical chorioamnionitis diagnosed during labor or maternal temperature $\geq$38 C (100.4 F)
- □ Moderate/heavy meconium staining of the amniotic fluid
- □ Fetal intolerance of labor such that one or more of the following actions was taken: in-utero resuscitative measures, further fetal assessment, or operative delivery
- □ Epidural or spinal anesthesia during labor
- □ None of the above

46. METHOD OF DELIVERY
- A. Was delivery with forceps attempted but unsuccessful? □ Yes □ No
- B. Was delivery with vacuum extraction attempted but unsuccessful? □ Yes □ No
- C. Fetal presentation at birth
  - □ Cephalic
  - □ Breech
  - □ Other
- D. Final route and method of delivery (Check one)
  - □ Vaginal/Spontaneous
  - □ Vaginal/Forceps
  - □ Vaginal/Vacuum
  - □ Cesarean
    If cesarean, was a trial of labor attempted? □ Yes □ No

47. MATERNAL MORBIDITY (Check all that apply) (Complications associated with labor and delivery)
- □ Maternal transfusion
- □ Third or fourth degree perineal laceration
- □ Ruptured uterus
- □ Unplanned hysterectomy
- □ Admission to intensive care unit
- □ Unplanned operating room procedure following delivery
- □ None of the above

NEWBORN INFORMATION

NEWBORN

48. NEWBORN MEDICAL RECORD NUMBER:
49. BIRTHWEIGHT (grams preferred, specify unit) ______ □ grams □ lb/oz
50. OBSTETRIC ESTIMATE OF GESTATION: ______ (completed weeks)
51. APGAR SCORE:
Score at 5 minutes: ______
If 5 minute score is less than 6,
Score at 10 minutes: ______
52. PLURALITY - Single, Twin, Triplet, etc. (Specify)______
53. IF NOT SINGLE BIRTH - Born First, Second, Third, etc. (Specify)______

54. ABNORMAL CONDITIONS OF THE NEWBORN (Check all that apply)
- □ Assisted ventilation required immediately following delivery
- □ Assisted ventilation required for more than six hours
- □ NICU admission
- □ Newborn given surfactant replacement therapy
- □ Antibiotics received by the newborn for suspected neonatal sepsis
- □ Seizure or serious neurologic dysfunction
- □ Significant birth injury (skeletal fracture(s), peripheral nerve injury, and/or soft tissue/solid organ hemorrhage which requires intervention)
- □ None of the above

55. CONGENITAL ANOMALIES OF THE NEWBORN (Check all that apply)
- □ Anencephaly
- □ Meningomyelocele/Spina bifida
- □ Cyanotic congenital heart disease
- □ Congenital diaphragmatic hernia
- □ Omphalocele
- □ Gastroschisis
- □ Limb reduction defect (excluding congenital amputation and dwarfing syndromes)
- □ Cleft Lip with or without Cleft Palate
- □ Cleft Palate alone
- □ Down Syndrome
  - □ Karyotype confirmed
  - □ Karyotype pending
- □ Suspected chromosomal disorder
  - □ Karyotype confirmed
  - □ Karyotype pending
- □ Hypospadias
- □ None of the anomalies listed above

Mother s Name______ Mother s Medical Record No.______

56. WAS INFANT TRANSFERRED WITHIN 24 HOURS OF DELIVERY? □ Yes □ No
IF YES, NAME OF FACILITY INFANT TRANSFERRED TO:______
57. IS INFANT LIVING AT TIME OF REPORT? □ Yes □ No □ Infant transferred, status unknown
58. IS THE INFANT BEING BREASTFED AT DISCHARGE? □ Yes □ No

REV. 11/2003

NOTE: This recommended standard birth certificate is the result of an extensive evaluation process. Information on the process and resulting recommendations as well as plans for future activities is available on the Internet at: http://www.cdc.gov/nchs/vital_certs_rev.htm.

Figure 4.3 Standard Birth and Death Certificates Used in the United States

U.S. STANDARD CERTIFICATE OF DEATH

LOCAL FILE NO. STATE FILE NO.

NAME OF DECEDENT ____________
For use by physician or institution

To Be Completed/Verified By: FUNERAL DIRECTOR

1. DECEDENT S LEGAL NAME (Include AKA s if any) (First, Middle, Last)
2. SEX
3. SOCIAL SECURITY NUMBER

4a. AGE-Last Birthday (Years)
4b. UNDER 1 YEAR — Months / Days
4c. UNDER 1 DAY — Hours / Minutes
5. DATE OF BIRTH (Mo/Day/Yr)
6. BIRTHPLACE (City and State or Foreign Country)

7a. RESIDENCE-STATE
7b. COUNTY
7c. CITY OR TOWN
7d. STREET AND NUMBER
7e. APT. NO.
7f. ZIP CODE
7g. INSIDE CITY LIMITS? ☐ Yes ☐ No

8. EVER IN US ARMED FORCES? ☐ Yes ☐ No
9. MARITAL STATUS AT TIME OF DEATH ☐ Married ☐ Married, but separated ☐ Widowed ☐ Divorced ☐ Never Married ☐ Unknown
10. SURVIVING SPOUSE S NAME (If wife, give name prior to first marriage)

11. FATHER S NAME (First, Middle, Last)
12. MOTHER S NAME PRIOR TO FIRST MARRIAGE (First, Middle, Last)

13a. INFORMANT S NAME
13b. RELATIONSHIP TO DECEDENT
13c. MAILING ADDRESS (Street and Number, City, State, Zip Code)

14. PLACE OF DEATH (Check only one: see instructions)
IF DEATH OCCURRED IN A HOSPITAL: ☐ Inpatient ☐ Emergency Room/Outpatient ☐ Dead on Arrival
IF DEATH OCCURRED SOMEWHERE OTHER THAN A HOSPITAL: ☐ Hospice facility ☐ Nursing home/Long term care facility ☐ Decedent s home ☐ Other (Specify):

15. FACILITY NAME (If not institution, give street & number)
16. CITY OR TOWN, STATE, AND ZIP CODE
17. COUNTY OF DEATH

18. METHOD OF DISPOSITION: ☐ Burial ☐ Cremation ☐ Donation ☐ Entombment ☐ Removal from State ☐ Other (Specify): ____________
19. PLACE OF DISPOSITION (Name of cemetery, crematory, other place)

20. LOCATION-CITY, TOWN, AND STATE
21. NAME AND COMPLETE ADDRESS OF FUNERAL FACILITY

22. SIGNATURE OF FUNERAL SERVICE LICENSEE OR OTHER AGENT
23. LICENSE NUMBER (Of Licensee)

To Be Completed By: MEDICAL CERTIFIER

ITEMS 24-28 MUST BE COMPLETED BY PERSON WHO PRONOUNCES OR CERTIFIES DEATH
24. DATE PRONOUNCED DEAD (Mo/Day/Yr)
25. TIME PRONOUNCED DEAD

26. SIGNATURE OF PERSON PRONOUNCING DEATH (Only when applicable)
27. LICENSE NUMBER
28. DATE SIGNED (Mo/Day/Yr)

29. ACTUAL OR PRESUMED DATE OF DEATH (Mo/Day/Yr) (Spell Month)
30. ACTUAL OR PRESUMED TIME OF DEATH
31. WAS MEDICAL EXAMINER OR CORONER CONTACTED? ☐ Yes ☐ No

CAUSE OF DEATH (See instructions and examples)

32. **PART I.** Enter the chain of events--diseases, injuries, or complications--that directly caused the death. DO NOT enter terminal events such as cardiac arrest, respiratory arrest, or ventricular fibrillation without showing the etiology. DO NOT ABBREVIATE. Enter only one cause on a line. Add additional lines if necessary.

Approximate interval: Onset to death

IMMEDIATE CAUSE (Final disease or condition ---------> resulting in death) a.____________
Due to (or as a consequence of):
Sequentially list conditions, if any, leading to the cause listed on line a. Enter the **UNDERLYING CAUSE** (disease or injury that initiated the events resulting in death) **LAST**
b.____________
Due to (or as a consequence of):
c.____________
Due to (or as a consequence of):
d.____________

PART II. Enter other significant conditions contributing to death but not resulting in the underlying cause given in PART I.
33. WAS AN AUTOPSY PERFORMED? ☐ Yes ☐ No
34. WERE AUTOPSY FINDINGS AVAILABLE TO COMPLETE THE CAUSE OF DEATH? ☐ Yes ☐ No

35. DID TOBACCO USE CONTRIBUTE TO DEATH? ☐ Yes ☐ Probably ☐ No ☐ Unknown
36. IF FEMALE: ☐ Not pregnant within past year ☐ Pregnant at time of death ☐ Not pregnant, but pregnant within 42 days of death ☐ Not pregnant, but pregnant 43 days to 1 year before death ☐ Unknown if pregnant within the past year
37. MANNER OF DEATH ☐ Natural ☐ Homicide ☐ Accident ☐ Pending Investigation ☐ Suicide ☐ Could not be determined

38. DATE OF INJURY (Mo/Day/Yr) (Spell Month)
39. TIME OF INJURY
40. PLACE OF INJURY (e.g., Decedent s home; construction site; restaurant; wooded area)
41. INJURY AT WORK? ☐ Yes ☐ No

42. LOCATION OF INJURY: State: City or Town: Street & Number: Apartment No.: Zip Code:

43. DESCRIBE HOW INJURY OCCURRED:
44. IF TRANSPORTATION INJURY, SPECIFY: ☐ Driver/Operator ☐ Passenger ☐ Pedestrian ☐ Other (Specify)

45. CERTIFIER (Check only one):
☐ Certifying physician-To the best of my knowledge, death occurred due to the cause(s) and manner stated.
☐ Pronouncing & Certifying physician-To the best of my knowledge, death occurred at the time, date, and place, and due to the cause(s) and manner stated.
☐ Medical Examiner/Coroner-On the basis of examination, and/or investigation, in my opinion, death occurred at the time, date, and place, and due to the cause(s) and manner stated.
Signature of certifier:____________

46. NAME, ADDRESS, AND ZIP CODE OF PERSON COMPLETING CAUSE OF DEATH (Item 32)

47. TITLE OF CERTIFIER
48. LICENSE NUMBER
49. DATE CERTIFIED (Mo/Day/Yr)
50. **FOR REGISTRAR ONLY**- DATE FILED (Mo/Day/Yr)

To Be Completed By: FUNERAL DIRECTOR

51. DECEDENT S EDUCATION-Check the box that best describes the highest degree or level of school completed at the time of death.
☐ 8th grade or less
☐ 9th - 12th grade; no diploma
☐ High school graduate or GED completed
☐ Some college credit, but no degree
☐ Associate degree (e.g., AA, AS)
☐ Bachelor s degree (e.g., BA, AB, BS)
☐ Master s degree (e.g., MA, MS, MEng, MEd, MSW, MBA)
☐ Doctorate (e.g., PhD, EdD) or Professional degree (e.g., MD, DDS, DVM, LLB, JD)

52. DECEDENT OF HISPANIC ORIGIN? Check the box that best describes whether the decedent is Spanish/Hispanic/Latino. Check the No box if decedent is not Spanish/Hispanic/Latino.
☐ No, not Spanish/Hispanic/Latino
☐ Yes, Mexican, Mexican American, Chicano
☐ Yes, Puerto Rican
☐ Yes, Cuban
☐ Yes, other Spanish/Hispanic/Latino (Specify) ____________

53. DECEDENT S RACE (Check one or more races to indicate what the decedent considered himself or herself to be)
☐ White
☐ Black or African American
☐ American Indian or Alaska Native (Name of the enrolled or principal tribe) ____________
☐ Asian Indian
☐ Chinese
☐ Filipino
☐ Japanese
☐ Korean
☐ Vietnamese
☐ Other Asian (Specify)____________
☐ Native Hawaiian
☐ Guamanian or Chamorro
☐ Samoan
☐ Other Pacific Islander (Specify)____________
☐ Other (Specify)____________

54. DECEDENT S USUAL OCCUPATION (Indicate type of work done during most of working life. DO NOT USE RETIRED).

55. KIND OF BUSINESS/INDUSTRY

REV. 11/2003

Figure 4.3 *(continued)*

Initially, the information collected about deaths indicated only the cause (since one goal was to keep track of the deadly plague), but starting in the eighteenth century the age of those dying was also noted. Yet despite the interest in these data created by the analyses of Graunt, Petty, Halley, and others mentioned in Chapter 3, people remained skeptical about the quality of the data and unsure of what could be done with them (Starr 1987). So it was not until the middle of the nineteenth century that civil registration of births and deaths became compulsory and an office of vital statistics was officially established by the English government, mirroring events in much of Europe and North America. It was not until 1900 that birth and death certificates were standardized in the United States.

Today, we find the most complete vital registration systems in the most highly developed countries and the least complete (often nonexistent) systems in the least developed countries. Such systems seem to be tied to literacy (there must be someone in each area to record events), adequate communication, and the cost of the bureaucracy required for such record keeping, all of which is associated with economic development. Among countries where systems of vital registration do exist, there is wide variation in the completeness with which events are recorded. Even in the United States, the registration of births is not 100 percent complete, yet the public so takes for granted the existence of vital statistics that the National Research Council was asked to convene a panel in 2009 to lay out the case for why continued funding is so vital (no pun intended) to our knowledge about the health of the nation (Siri and Cork 2009).

Although most nations have a system of birth and death registration that is separate from census activities, dozens of countries, mostly in Europe, maintain **population registers**, which are lists of all people in the country, and which can be used as a substitute for a census, as I mentioned earlier. Alongside each name are recorded the vital events for that individual, typically birth, death, marriage, divorce, and change of residence. Such registers are kept primarily for administrative (that is, social control) purposes, such as legal identification of people, election rolls, and calls for military service, but they are also extremely valuable for demographic purposes, since they provide a demographic life history for each individual. Even though registers are expensive to maintain, many countries that could afford them, such as the United States, tend to avoid them because of the perceived threat to personal freedom that can be inherent in a system that compiles and centralizes personally identifying information.

Combining the Census and Vital Statistics

Although recording vital events provides information about the number of births and deaths (along with other events) according to such characteristics as age and sex, we also need to know how many people are at risk of these events. Thus, vital statistics data are typically teamed up with census data, which do include that information. For example, you may know from the vital statistics that there were 4.0 million births in the United States in 2012 (the most recent year for which data are available at this writing), but that number tells you nothing about whether the birth rate was high or low. In order to draw any conclusion, you must relate those births to the 314 million people residing in the United States as of mid-2012, and only then do you discover a relatively low birth rate of 12.7 births per 1,000 population, down from 16.7 in 1990.

Since in 2012 no census had been taken since 2010, you may wonder how an estimate of the population could have been produced for an **intercensal** year such as 2012. The answer is that once again census data are combined with vital statistics data (and migration estimates) using the demographic balancing equation that I discussed earlier in the chapter: the population in 2012 is equal to the population as of the 2010 census, plus the births, minus the deaths, plus the in-migrants, minus the out-migrants between 2010 and 2012. Naturally, deficiencies in any of these data sources will lead to inaccuracies in the estimate of the number of people alive at any time, but we typically won't know that until we conduct the next census.

Administrative Data

Knowing that censuses and the collection of vital statistics were not originally designed to provide data for demographic analysis has alerted demographers everywhere to keep their collective eyes open for any data source that might yield useful information. For example, an important source of information about immigration to the United States is the compilation of **administrative records** filled out for each person entering the country from abroad. These forms are collected and tabulated by the U.S. Citizenship and Immigration Service (USCIS) within the U.S. Department of Homeland Security. Of course, we need other means to estimate the number of people who enter without documents and avoid detection by the government, and I discuss that more in Chapter 7.

Data are not routinely gathered on people who permanently leave the United States, but the administrative records of the U.S. Social Security Administration provide some clues about the number and destination of such individuals because many people who leave the country have their Social Security checks follow them. An administrative source of information on migration within the United States used by the Census Bureau is a set of data provided to them by the Internal Revenue Service (IRS). Although no personal information is ever divulged, the IRS can match Social Security numbers of taxpayers each year and see if their address has changed, thus providing a clue about geographic mobility, at least among those people who file income tax returns. At the local level, a variety of administrative data can be tapped to determine demographic patterns. School enrollment data provide clues to patterns of population growth and migration. Utility data on connections and disconnections can also be used to discern local population trends, as can the number of people signing up for government-sponsored health programs (Medicaid and Medicare) and income assistance (various forms of welfare).

Sample Surveys

There are two major difficulties with using data collected in the census, by the vital statistics registration system, or derived from administrative records: (1) They are usually collected for purposes other than demographic analysis and thus do not necessarily reflect the theoretical concerns of demography; and (2) they are collected by many different people using many different methods and may be prone to numerous kinds of error. For these two reasons, in addition to the cost of big data–collection schemes, sample surveys are frequently used to gather demographic data. Sample surveys may provide the social, psychological, economic, and even physical data I referred to earlier as being necessary to an understanding of why things are as they are. Their principal limitation is that they provide less extensive geographic coverage than a census or system of vital registration.

By using a carefully selected sample of even a few thousand people, demographers have been able to ask questions about births, deaths, migration, and other subjects that reveal aspects of the "why" of demographic events rather than just the "what." In some poor or remote areas of the world, sample surveys can also provide good estimates of the levels of fertility, mortality, and migration in the absence of census or vital registration data.

Demographic Surveys in the United States

I have already mentioned the American Community Survey (ACS), which is now a critically important part of the census itself. It is modeled after the Current Population Survey (CPS) conducted monthly by the U.S. Census Bureau in collaboration with the Bureau of Labor Statistics, and which for many decades has been one of the country's most important sample surveys. Since 1943, thousands of households (currently more than 50,000) have been queried each month about a variety of things, although a major thrust of the survey is to gather information on the labor force. Each March, detailed demographic questions are also asked about fertility and migration and such characteristics as education, income, marital status, and living arrangements. These data have been an important source of demographic information about the U.S. population, filling in the gap between censuses, and providing the Census Bureau with the experience necessary to launch the more ambitious ACS.

Since 1983, the Census Bureau has also been conducting the Survey on Income and Program Participation (SIPP), which is a companion to the Current Population Survey. Using a rotating panel of more than 40,000 households that are queried several times over a two- to four-year period, the SIPP gathers detailed data on sources of income and wealth, disability, and the extent to which household members participate in government assistance programs. The Census Bureau also regularly conducts the American Housing Survey for the U.S. Department of Housing and Urban Development, and this survey generates important data on mobility and migration patterns in the United States. The National Center for Health Statistics (NCHS) within the U.S. Centers for Disease

Control and Prevention (CDC) generates data about fertility and reproductive health in the National Survey of Family Growth (NSFG), which it conducts every five years or so, and also obtains data on health and disability from the annual National Health Interview Survey (NHIS). These latter data are now available for each year from 1963 to the present from the Minnesota Population Center's IPUMS website.

Canadian Surveys

Canada has a monthly Labour Force Survey (LFS), initiated in 1945 to track employment trends after the end of World War II. Similar to the CPS in the United States, it is a rotating panel of 56,000 households, and although its major purpose is to produce data on the labor force (hence the name), it gathers data on most of the core sociodemographic characteristics of people in each sampled household, so it provides a continuous measure of population trends in Canada.

Since 1985, Statistics Canada has also conducted an annual General Social Survey, a sample of about 25,000 respondents. Each survey has a different set of in-depth topics designed to elicit detailed data about various aspects of life in Canada, such as health and social support, families, and time use. Like the National Household Survey, this is a voluntary survey and as the response rate from the random digit dialing sample dropped below 65 percent in 2010, an online questionnaire was added in order to boost participation (Statistics Canada 2013b).

Mexican Surveys

Mexico conducts several regular national household surveys, one of which in particular is comparable to the CPS and the LFS. The National Survey of Occupation and Employment (Encuesta Nacional de Ocupación y Empleo [ENOE]) is a large (120,000 household) sample of households undertaken three times a year by INEGI and is designed to be representative of the entire country. As with the CPS and LFS, the goal is to provide a way of regularly measuring and monitoring the social and economic characteristics of the population beyond just data on current employment. Some of the population questions asked in the census (see Table 4.1) are also asked in the ENOE, along with a detailed set of questions about the labor force activity of everyone in the household who is 12 years of age or older.

Demographic and Health Surveys

As I noted above, most developing countries do not have good systems of vital registration, without which it is difficult to track changes in mortality and fertility. Into this breach have stepped the Demographic and Health Surveys (DHS). This is the largest and globally most important set of demographic surveys and they are technically part of the Measure DHS project of ICF International in Maryland, conducted with funding from the U.S. Agency for International Development (USAID).

The DHS is actually the successor to the World Fertility Survey, which was conducted between 1972 and 1982 under the auspices of the International Statistical Institute in the Netherlands. Concurrent with the World Fertility Survey was a series of Contraceptive Prevalence Surveys, conducted in Latin America, Asia, and Africa with funding from the U.S. Agency for International Development (USAID). In 1984, the work of the World Fertility Survey and the Contraceptive Prevalence Surveys was combined into the Demographic and Health Surveys. The focus is on fertility, reproductive health, and child health and nutrition, but the data provide national estimates of basic demographic processes, structure, and characteristics, since a few questions are asked about all members of each household in the sample. More than 300 surveys have been conducted in more than 90 developing countries in Africa, Asia, and Latin America. This is a rich source of information, as you will see in subsequent chapters.

A complementary set of surveys has been conducted in poorer countries that, for a variety of reasons, have not had a Demographic and Health Survey. Known as the Multiple Indicator Cluster Surveys (MICS), they were developed by the United Nations Children's Fund (UNICEF) and are funded by a variety of international agencies. These surveys collect data that are similar to those in the DHS.

Demographic Surveillance Systems

In Africa, many people are born, live, and die without a single written record of their existence because of the poor coverage of censuses and vital registration systems. The INDEPTH Network was created in 1998 to provide a way of tracking the lives of people in specific "sentinel" areas of sub-Saharan Africa (and to a lesser extent south Asia) by working with individual countries to select one or two defined geographic regions that are representative of a larger population. A census is conducted in that region, and then subsequent demographic changes are continuously measured by keeping track of all births, deaths, migration, and related characteristics of the population. There are currently 42 surveillance sites in 20 different countries of Africa and Asia. INDEPTH was funded initially by governmental organizations, especially the Canadian government, and is now funded largely through private foundations, including the Bill and Melinda Gates Foundation.

European Surveys

Declining fertility and the concomitant aging of the population in Europe has generated a renewed interest in the continent's demography, and there are now several surveys in Europe that capture useful demographic information. The Population Unit of the United Nations Economic Commission for Europe funded the Family and Fertility Surveys (FFS) in 23 European nations during the 1990s. Since 2000, they have funded the "Generations and Gender Program," which is a longitudinal survey of 18 to 79-year-olds in 19 countries gathering data on a broad array of topics including fertility, partnership, the transition to adulthood, economic activity, care duties and attitudes.

The European Social Survey (ESS) is a cross-national survey that has been conducted every two years across Europe since 2001 by researchers at City University London. The survey measures attitudes, beliefs and behavior patterns, along with the demographics of populations in more than thirty European nations. It is funded by the European Commission and the European Science Foundation.

Historical Sources

Our understanding of population processes is shaped not only by our perception of current trends but also by our understanding of historical events. Historical demography requires that we almost literally dig up information about the patterns of mortality, fertility, and migration in past generations—to reconstruct "the world we have lost," as Peter Laslett (1971) once called it. You may prefer to whistle past the graveyard, but researchers at the Cambridge Group for the History of Population and Social Structure in the Department of Geography and the Faculty of History at Cambridge University (U.K.) have spent the past several decades developing ways to recreate history by reading dates on tombstones and organizing information contained in parish church registers and other local documents (Wrigley and Schofield 1981; Reher and Schofield 1993), extending methods developed especially by the great French historical demographer Louis Henry (1967; Rosental 2003).

Historical sources of demographic information include censuses and vital statistics, but the general lack of good historical vital statistics is what typically necessitates special detective work to locate birth records in church registers and death records in graveyards. Even in the absence of a census, a complete set of good local records for a small village may allow a researcher to reconstruct the demographic profile of families by matching entries of births, marriages, and deaths in the community over a period of several years. Yet another source of such information is family genealogies, the compilation of which has become increasingly common in recent years throughout the world. Detailed genealogies in China, for example, have allowed researchers at Cambridge University to develop simulation models of what the demographic structure of China must have been like in the past (Zhao 2001).

The results of these labors can be of considerable importance in testing our notions about how the world used to work. For example, through historical demographic research we now know that the conjugal family (parents and their children) is not a product of industrialization and urbanization, as was once thought (Wrigley 1974). In fact, such small family units were quite common throughout Europe for several centuries before the Industrial Revolution and may actually have contributed to the process of industrialization by allowing the family more flexibility to meet the needs of the changing economy. In subsequent chapters, we will also have numerous occasions to draw on the results of the Princeton European Fertility Project, which gathered and analyzed data on marriage and reproduction throughout nineteenth- and early-twentieth-century Europe, as I discussed in Chapter 3.

By quantifying (and thereby clarifying) our knowledge of past patterns of demographic events, we are also better able to interpret historical events in a meaningful fashion. In the United States extended families may have been more common prior to the nineteenth century than has generally been thought (Ruggles 1994). Indeed, Wells (1982) has reminded us that the history of the struggle of American colonists to survive, marry, and bear children may tell us more about the determination to forge a union of states than does a detailed recounting of the actions of British officials.

Spatial Demography

Spatial demography represents the application of spatial concepts and statistics to demographic phenomena (Weeks 2004; Voss 2007; Matthews and Parker 2013). It recognizes that demography is, by its very nature, concerned with people in places. Since people tend to do things differently in different places, demography is inherently spatial. Where you live is an important determinant of who you are, and social scientists are increasingly aware that spatial variation is a universal principle of human society. For example, the innovation of the early fertility declines in Europe, which I discussed in Chapter 3, provides a nearly classic example of Waldo Tobler's First Law of Geography that everything is related to everything else, but near things are more related than distant things (Tobler 1970, 2004). This is a concept known as **spatial autocorrelation.** Thus, in Europe, had it not been for spatial autocorrelation, fertility might have declined in isolated settings, but the decline would not have spread as it did. It turns out that all three demographic processes—mortality, fertility, and migration—exhibit spatial autocorrelation.

Because culture underlies most aspects of demography, understanding why some places have different cultures than others helps us to understand spatially varying levels of mortality, fertility, and migration. Recognizing and studying this spatial variability has been greatly enhanced by the technologies and tools that are wrapped into the overall field of Geographic Information Science (GIScience). As a result of these new methods of analysis and of viewing the world, demography is evolving from being a primarily spatially *aware* science (which it has always been) to an increasingly more spatially *analytic* science (facilitated by the methods of GIScience). The advent of high-powered personal computers has revolutionized our ability to analyze massive demographic data sets, and this has allowed the spatial component of demographic analysis to come into its own and further improve our knowledge of how the world works. The first uses of these concepts and methods actually occurred in business and government planning, and then migrated, if you will, to academic research. Let me provide you with one of the best examples of this from cluster marketing.

We often talk about the numbers and characteristics of people (their "demographic") in terms of the likelihood that they will buy certain kinds of products, watch certain kinds of movies, or vote for particular candidates. But where are those people? Where should you concentrate your resources in order to get their attention. The fact that "birds of a feather flock together" (i.e., that spatial autocorrelation is a regular feature of the world) means that neighborhoods can be identified on the basis of a whole set of shared sociodemographic characteristics. This greatly facilitates the process of marketing to particular groups in a process known as cluster marketing. This takes us back to the 1970s when . . .

> . . . a computer scientist turned entrepreneur named Jonathan Robbin devised a wildly popular target-marketing system by matching zip codes with census data and consumer surveys. Christening his creation PRIZM (Potential Rating Index for Zip Markets), he programmed computers to sort the nation's 36,000 zips into forty "lifestyle clusters." Zip 85254 in Northeast Phoenix, Arizona, for instance, belongs to what he called the Furs and Station Wagons cluster, where surveys indicate that residents tend to buy lots of vermouth, belong to a country club, read *Gourmet* and vote the GOP ticket. In 02151, a Revere Beach, Massachusetts, zip designated Old Yankee Rows, tastes lean toward beer, fraternal clubs, Lakeland Boating and whoever the Democrats are supporting. (Weiss 1988:xii)

The PRIZM system made Robbin's company, Claritas Corporation (now part of Nielsen), one of the largest and most successful spatial demographics (*aka* **geodemographics**) firms in the world. A core principle is that *where* you live is a good predictor of *how* you live (Weiss 2000; Harris et al. 2005). It combines demographic characteristics with lifestyle variables and permits a business to home in on the specific neighborhoods where its products can be most profitably marketed. In keeping with the changing demographics of America, Neilsen Claritas adds new clusters as neighborhoods evolve.

Mapping Demographic Data

Demographers have been using maps as a tool for analysis for a long time, and some of the earliest analyses of disease and death relied heavily on maps that showed, for example, where people were dying from particular causes. In the middle of the nineteenth century, London physician John Snow used maps to trace a local cholera epidemic. In research that established the modern field of epidemiology, Snow was able to show that cholera occurred much more frequently among customers of a water company that drew its water from the lower Thames River (downstream from the city), where it had become contaminated with London sewage. However, neighborhoods drawing water from another company were associated with far fewer cases of cholera because that company obtained water from the upper Thames—prior to its passing through London, before sewage was dumped in the river (Snow 1936).

Today a far more sophisticated version of this same idea is available to demographers through **geographic information systems (GIS)**; which form the major part of the field of GIScience. A GIS is a computer-based system that allows us to combine maps with data that refer to particular places on those maps and then to analyze those data using spatial statistics (part of GIScience) and display the results as thematic maps or some other graphic format. The computer allows us to transform a map into a set of areas (such as a country, state, or census tract), lines (such as streets, highways, or rivers), and points (such as a house, school, or a health clinic). Our demographic data must then be **geo-referenced** (associated with some geographic identification such as precise latitude-longitude coordinates, a street address, ZIP code, census tract, county, state, or country) so the computer will link them to the correct area, line, or point. Demographic data are virtually always referenced to a geographic area, and in the United States the Geography Division of the U.S. Census Bureau works closely with the Population Division to make sure that data are identified for appropriate levels of "census geography," as shown in Figure 4.4.

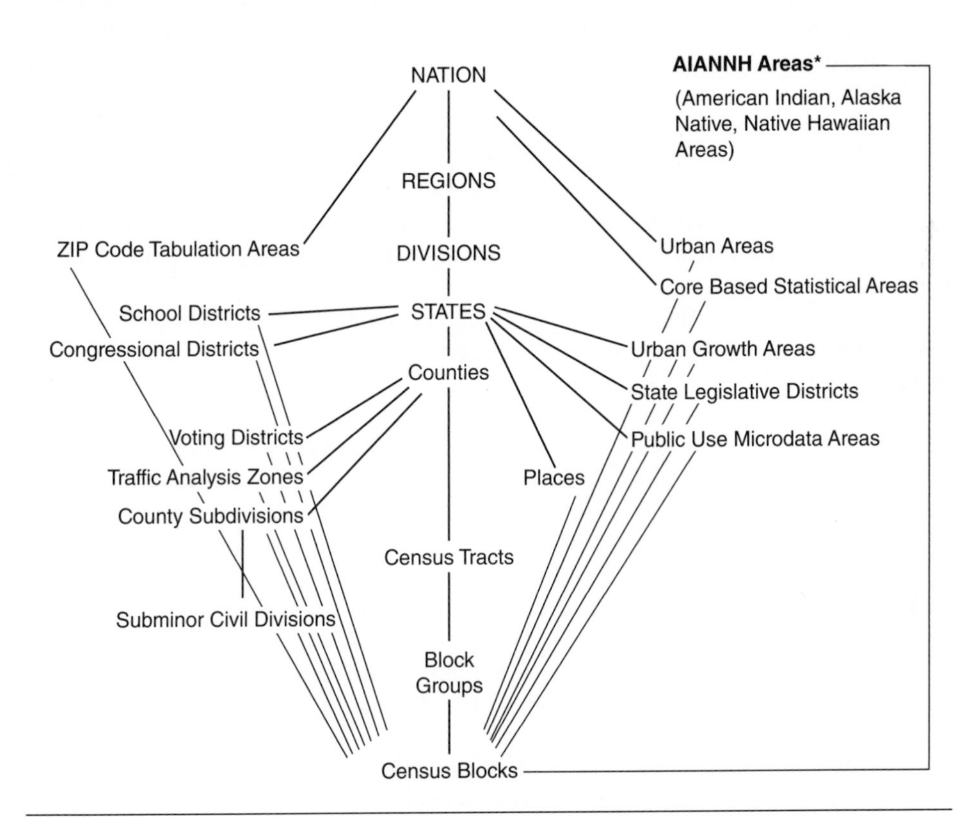

Figure 4.4 The U.S. Census Provides Geographically Referenced Data for a Wide Range of Geographic Areas

Source: (U.S. Census Bureau 2010)

Geo-referencing data to places on the map means we can combine different types of data (such as census and survey data) for the same place, and we can do it for more than one time (such as data for 2000 and 2010). Spatial demography thus improves and enhances our ability to visualize and analyze the kinds of demographic changes taking place over time and space. Since 1997, for example, most of the Demographic and Health Surveys in less developed countries have used global positioning system (GPS) devices (another geospatial technique) to record the location of (geo-reference) each household in the sample in order to allow for more sophisticated spatial demographic analysis of the survey data. An increasing number of surveys are doing the same thing, thus exponentially increasing our ability to understand demographic change.

GIS and the Census

It is a gross understatement to say that the computer has vastly expanded our capacity to process and analyze data. It is no coincidence that census data are so readily amenable to being "crunched" by the computer; the histories of the computer and the U.S. Census Bureau go back a long way together. Before the 1890 census, the U.S. government held a contest to see who could come up with the best machine for counting the data from that census. The winner was Herman Hollerith, who had worked on the 1880 census right after graduating from Columbia University. His method of feeding a punched card through a tabulating machine proved to be very successful, and in 1886 he organized the Tabulating Machine Company, which in 1911 was merged with two other companies and became the International Business Machines (IBM) Corporation (Kaplan and Van Valey 1980).

Then, after World War II, the Census Bureau sponsored the development of the first computer designed for mass data processing—the UNIVAC-I—which was used to help with the 1950 census and led the world into the computer age. Photo-optical scanning, which we now heavily rely on for entering data from printed documents into the computer (not to mention scanning the price of everything you buy at stores), was also a by-product of the Census Bureau's need for a device to tabulate data from census forms. FOSIDC (film optical sensing device for input to computers) was first used for the 1960 census.

Another useful innovation was the creation for the 1980 U.S. census of the DIME (Dual Independent Map Encoding) files. This was the first step toward computer mapping, in which each piece of data was coded in a way that could be matched electronically to a place on a map. In the 1980s, several private firms latched onto this technology, improved it, and made it available to other companies for their own business uses.

By the early 1990s, the pieces of the puzzle had come together. The data from the 1990 U.S. census were made available for the first time on CD-ROM and at prices affordable to a wide range of users. Furthermore, the Census Bureau reconfigured its geographic coding of data, creating what it calls TIGER

(Topologically Integrated Geographic Encoding and Referencing) files, which are digital boundary files that allow us to map the census data. At the same time, and certainly in response to increased demand, personal computers came along that were powerful enough and had enough memory to store and manipulate huge census files, including both the geographic database and the actual population and housing data. Not far behind was the software to run those computers, and several firms now make software for desktop computers that allow interactive spatial analysis of census and other kinds of data and then the production of high-quality color maps of the analysis results. Two of these firms—Environmental Systems Research Institute (ESRI) and Geographic Data Technology (GDT—now Tele Atlas)—have been working with the Census Bureau since before Census 2000 to help update the Census Bureau's computerized Master Address File (the information used to continuously update the TIGER files) in order to improve census coverage and geographic accuracy. In a very real sense, the census and the TIGER files, more specifically, helped to spawn the now-booming GIS industry. By the time the 2010 census rolled around, it had become possible to map census data online through the Census Bureau's website and to download digital boundary files for use on your own computer.

Knowledge and understanding are based on information, and our information base grows by being able to tap more deeply into rich data sources such as censuses and surveys. GIS is an effective tool for doing this, and you will see numerous examples of GIS at work in the remaining chapters. You can also see it at work on the Internet. Virtually all of the data from recent censuses and the American Community Survey are available on the U.S. Census Bureau's website, where you can create thematic maps on the fly.

Summary and Conclusion

The working bases of any science are facts and theory. In this chapter, I have discussed the major sources of demographic information, the wells from which population data are drawn. Censuses are the most widely known and used sources of data on populations, and humans have been counting themselves in this way for a long time. However, the modern series of more scientific censuses dates only from the late eighteenth and early nineteenth centuries. The high cost of censuses, combined with the increasing knowledge we have about the value of surveys, has meant that even so-called complete enumerations often include some kind of sampling. That is certainly true in North America, as the United States, Canada, and Mexico all use sampling techniques in their censuses. Even vital statistics can be estimated using sample surveys, especially in developing countries, although the usual pattern is for births and deaths (and often marriages, divorces, and abortions) to be registered with the civil authorities. Some countries take this a step further and maintain a complete register of life events for everybody.

Knowledge can also be gleaned from administrative data gathered for nondemographic purposes. These are particularly important in helping us measure migration. It is not just the present that we attempt to measure; historical sources of information can add much to our understanding of current trends in population growth and change. Our ability to know how the world works is increasingly enhanced by incorporating our demographic data into a geographic information system, permitting us to ask questions that were not really answerable before the advent of the computer. Spatial demography expands our demographic perspective into a geographic realm about which demographers have long been aware, but only recently have been able to analyze.

In this and the preceding three chapters, I have laid out for you the basic elements of a demographic perspective. With this in hand (and hopefully in your head as well), we are now ready to probe more deeply into the analysis of population processes; to come to an appreciation of how important the decline in the death rate is, yet why it is still so much higher in some places than in others, why birth rates are still high in some places yet very low in others, and why some people move and others do not.

Main Points

1. In order to study population processes and change, you need to know how many people are alive, how many are being born, how many are dying, how many are moving in and out, and why these things are happening.
2. A basic source of demographic information is the population census, in which information is obtained about all people in a given area at a specific time.
3. Not all countries regularly conduct censuses, but most of the population of the world has been enumerated since 2000.
4. Errors in the census typically come about as a result of nonsampling errors (the most important source of error, including coverage error and content error) or sampling errors.
5. It has been said that censuses are important because if you aren't counted, you don't count.
6. Information about births and deaths usually comes from vital registration records—data recorded and compiled by government agencies. The most complete vital registration systems are found in the most highly developed nations, while they are often nonexistent in less developed areas.
7. Most of the estimates of the magnitude of population growth and change are derived by combining census data with vital registration data (as well as administrative data), using the demographic balancing equation.
8. Sample surveys are sources of information for places in which census or vital registration data do not exist or where reliable information can be obtained less expensively by sampling than by conducting a census.

9. Parish records and old census data are important sources of historical information about population changes in the past.
10. Spatial demography involves using geographic information systems to analyze demographic data from a spatial perspective, thus contributing substantially to our understanding of how the world works.

Questions for Review:

1. In the United States, data are already collected from nearly everyone for Social Security cards and drivers' licenses. Why then does the country not have a population register that would eliminate the need for the census?
2. Survey data are never available at the same geographic detail as are census data. What are the disadvantages associated with demographic data that are not provided at a fine geographic scale?
3. Virtually all of the demographic surveys and surveillance systems administered in developing countries are paid for by governments in richer countries. What is the advantage to richer countries of helping less-rich countries to collect demographic data?
4. What is the value to us in the twenty-first century of having an accurate demographic picture of earlier centuries?
5. Provide an example of spatial autocorrelation from your own personal experience. How might this concept influence your demographic perspective?

Websites of Interest

Remember that websites are not as permanent as books and journals, so I cannot guarantee that each of the following websites still exists at the moment you are reading this. You may have to Google the name of the organization to find the current web address.

1. **http://unstats.un.org/unsd/demographic/sources/census/censusdates.htm**
   The United Nations Statistics Division facilitates census-taking throughout the world, and at this site you can see the current status of censuses undertaken or planned for each country.
2. **http://www.census.gov**
   The home page of the U.S. Census Bureau. From here you can locate an amazing variety of information, including the latest releases of census data, the American Community Survey, and all of the surveys conducted by the Census Bureau. This is one of the most accessed websites in the world.
3. **http://www.statcan.gc.ca**
   The home page of Statistics Canada, the government organization that conducts the censuses and surveys in Canada. From here you can obtain census data and track other demographically related information about Canada, including vital statistics and survey data, and you can do so in either English or French.

4. **http://www.inegi.gob.mx**
   The home page of INEGI (Instituto Nacional de Estadística, Geografía, y Informática), which is the government agency in Mexico that conducts the censuses and related demographic surveys, as well as compiling the vital statistics for Mexico. You can obtain all of the latest census and survey information from this site, although you will need to be able to read Spanish to do so.

5. **http://weekspopulation.blogspot.com/search/label/demographic%20data**
   Keep track of the latest news related to this chapter by visiting my WeeksPopulation website.

CHAPTER 5
The Health and Mortality Transition

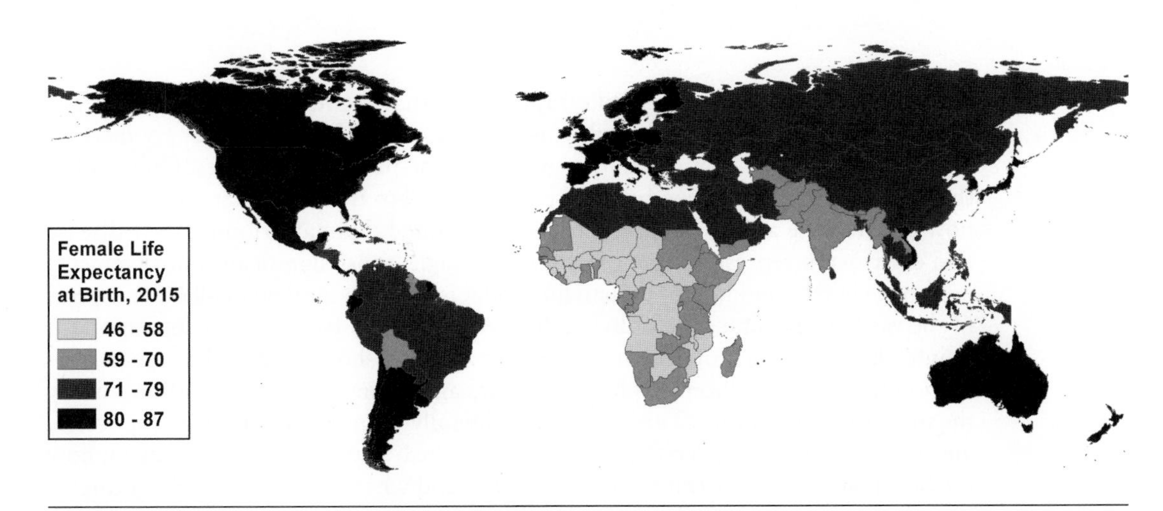

Figure 5.1 Global Variability in Female Life Expectancy at Birth

Source: Adapted by the author from data in United Nations Population Division (United Nations Population Division 2013).

DEFINING THE HEALTH AND MORTALITY TRANSITION

HEALTH AND MORTALITY CHANGES OVER TIME
The Roman Era to the Industrial Revolution
The Industrial Revolution to the Twentieth Century
World War II as a Modern Turning Point
Postponing Death by Preventing and Curing Disease
The Nutrition Transition and Its Link to Obesity

LIFE SPAN AND LONGEVITY
Life Span
Longevity

DISEASE AND DEATH OVER THE LIFE CYCLE
Age Differentials in Mortality
Infant Mortality
Mortality at Older Ages
Sex and Gender Differentials in Mortality

CAUSES OF POOR HEALTH AND DEATH
Communicable Diseases
HIV/AIDS
Emerging Infectious Diseases
Maternal Mortality
Noncommunicable Conditions
Injuries
The "Real" Causes of Death

MEASURING MORTALITY
Crude Death Rate
Age/Sex-Specific Death Rates
Age-Adjusted Death Rates
Life Tables
Life Table Calculations
Disability-Adjusted Life Years

HEALTH AND MORTALITY INEQUALITIES
Urban and Rural Differentials
Neighborhood Inequalities
Educational and Socioeconomic Differentials in Mortality
Inequalities by Race and Ethnicity
Marital Status and Mortality

ESSAY: Mortality Control and the Environment

It isn't that people now breed like rabbits; it's that we no longer die like flies—declining mortality, not rising fertility, is the root cause of the revolutionary increase in the world's population size and growth over the past two centuries. Only within that time has mortality been brought under control to the point that most of us are now able to take a long life pretty much for granted. Human triumph over disease and early death represents one of, if not the single, most significant improvements ever made in the condition of human life, and it is tightly bound up in all other aspects of the vastly higher standard of living that we now enjoy. Nevertheless, an important unintended by-product of declining mortality is the mushrooming of the human population from just 1 billion two hundred years ago to an expected 9 to 10 billion by the middle of this century. This increase has literally changed everything in the world, and you cannot fully understand the world in which you live without knowing how the health and mortality transition came about and what this means for the future.

I begin the chapter with a brief description of the health and mortality transition and then illustrate its impact by reviewing the changes in health and mortality over time, up to the present. The transition is by no means over, however, so I next consider how far it can go, given what we know about human life span and longevity, and about the things that can and do kill us and what we are doing about them. We will measure the progress of the transition using a variety of indices that I review in the chapter, and I employ some of those tools in the last part of the chapter to examine important inequalities that exist in the world with respect to health and mortality.

Defining the Health and Mortality Transition

Health and death are typically thought of as two sides of the same coin—morbidity and mortality, respectively—with morbidity referring to the prevalence of disease in a population and mortality the pattern of death. The link is a familiar one to most people—the healthier you are, the longer you are likely to live. At the societal level, this means that populations with high morbidity are those with high mortality; therefore, as health levels improve, so does life expectancy. Most of us in the richer countries take our long life expectancy for granted. Yet scarcely a century ago, and for virtually all of human history before that, death rates were very high and early death was commonplace. Within the past 200 years, and especially during the twentieth century, country after country has experienced a transition to better health and lower death rates—a long-term shift in health and disease patterns that has brought death rates down from very high levels in which people die young, primarily from communicable diseases, to low levels, with deaths concentrated among the elderly, who die from degenerative diseases. This phenomenon was originally defined by Abdel Omran (1971, 1977) as the "epidemiological transition," but since the term "epidemiology" technically refers only to disease and not to death, I have chosen to broaden the term to the health and mortality transition.

As a result of this transition, the variability by age in mortality is reduced or *compressed*, leading to an increased rectangularization of mortality. This means that most people survive to advanced ages and then die pretty quickly (as I will discuss in more detail later in the chapter). The vast changes in society brought about as more people survive to ever older ages represent important contributions to the overall demographic transition. We can begin to understand this most readily by examining how health and mortality have changed dramatically during the course of human history, especially European history, for which we tend to have better data than for the rest of the world.

Health and Mortality Changes Over Time

In much of the world and for most of human history, life expectancy probably fluctuated between 20 and 30 years (United Nations 1973; Weiss 1973; Riley 2005). At this level of mortality, only about two-thirds of babies survived to their first birthday, and only about one-half were still alive at age five, as seen in Table 5.1. This means that one-half of all deaths occurred before age five. At the other end of

Table 5.1 The Meaning of Improvements in Life Expectancy

Period	Life Expectancy for Females	Percentage Surviving to Age:				Percentage of Deaths:		Number of Births Required for ZPG
		1	5	25	65	<5	65+	
Premodern	20	63	47	34	8	53	8	6.1
	30	74	61	50	17	39	17	4.2
US and Europe in late eighteenth and early nineteenth centuries	40	82	73	63	29	27	29	3.3
Lowest in sub-Saharan African	46	89	82	75	34	18	34	2.7
World average circa 2015	73	98	98	97	77	2	77	2.1
Mexico	78	99	99	98	84	1	84	2.1
United States	81	99	99	99	88	<1	88	2.1
Canada	84	99	99	99	91	<1	91	2.1
Japan (highest in world)	86	99	99	99	93	<1	93	2.1

Sources: Life expectancies less than 69 are based on stable population models in Ansley Coale and Paul Demeny, *Regional Model Life Tables and Stable Populations* (Princeton, NJ: Princeton University Press, 1966); other life table are from World Health Organization, Global Health Observatory Data Repository, http://apps.who.int/gho/data/node.main.692?lang=en, accessed 2014.

the age continuum, around 10 percent of people made it to age 65 in a premodern society. Thus, in the premodern world, about one-half the deaths were to children under age five and only about one in 10 were to a person aged 65 or older.

In hunter-gatherer societies, it is likely that the principal cause of death was poor nutrition—people literally starving to death—combined perhaps with selective infanticide and geronticide (the killing of older people) (McKeown 1988), although there is too little evidence to do more than speculate about this (Bocquet-Appel 2008). As humans gained more control over the environment by domesticating plants and animals (the Agricultural Revolution), both birth and death rates probably went up, as I mentioned in Chapter 2. It was perhaps in the sedentary, more densely settled villages common after the Agricultural Revolution that infectious diseases became a more prevalent cause of death. People were almost certainly better fed, but closer contact with one another, with animals, and with human and animal waste encouraged the spread of disease, with especially disastrous results for infants, a situation that prevailed for thousands of years.

The Roman Era to the Industrial Revolution

Life expectancy in the Roman era is estimated to have been 22 years (Petersen 1975). Keep in mind that this does not mean that everybody dropped dead at age 22. Looking at Table 5.1 you can see that it means the majority of children born did not survive to adulthood. People who reached adulthood were not too likely to reach a very advanced age, but of course some did. The major characteristic of high mortality societies was that there was a lot more variability in the ages at which people died than is true today, but in general people died at a younger, rather than an older, age.

The Roman empire began to break up by the third century, and the period from about the fifth to the fifteenth centuries represents the Middle Ages. Nutrition in Europe during this period probably improved enough to raise life expectancy to more than 30 years. The plague, or Black Death, hit Europe in the fourteenth century, having spread west from Asia (Cantor 2001; Christakos et al. 2005). It is estimated that one-third of the population of Europe may have perished from the disease between 1346 and 1350. The plague then made a home for itself in Europe and, as Cipolla says, "For more than three centuries epidemics of plague kept flaring up in one area after another. The recurrent outbreaks of the disease deeply affected European life at all levels—the demographic as well as the economic, the social as well as the political, the artistic as well as the religious" (Cipolla 1981:3). The constant uncertainty about life could crush you, but it could also encourage you to take risks, as some Europeans did, spreading out around the world.

I mentioned in Chapter 2 that Europe's increasing dominance in oceanic shipping and weapons gave it an unrivaled ability not only to trade goods with the rest of the world but to trade diseases as well. The most famous of these disease transfers was the so-called **Columbian Exchange**, involving the diseases that Columbus and other European explorers took to the Americas (and a few that they took back to Europe). Their relative immunity to the diseases they brought with them, at least in comparison with the devastation those diseases wrought on the indigenous populations, is one explanation for the relative ease with which Spain was able to dominate much of Latin America after arriving there around 1500. The populations in Middle America at the time of European conquest were already living under conditions of "severe nutritional stress and extremely high mortality" (McCaa 1994:7), but contact with the Spaniards turned a bad situation into what Robert McCaa (1994) has called a "demographic hell," with high rates of orphanhood and with life expectancy probably dipping below 20 years. Spain itself was hit by at least three major plague outbreaks between 1596 and 1685, and William McNeill (1976) suggests that this may have been a significant factor in Spain's decline as an economic and political power.

The Industrial Revolution to the Twentieth Century

The plague had been more prevalent in the Mediterranean area (where it is too warm for the fleas to die during the winter) than farther north or east, and the last major sighting of the plague in Europe was in the south of France, in Marseilles, in 1720. It is no coincidence that this was the eve of the Industrial Revolution. The plague retreated (rather than disappeared) probably as a result of "changes in housing, shipping, sanitary practices, and similar factors affecting the way rats, fleas, and humans encountered one another" (McNeill 1976:174), and other causes of poor health were diminished by the receding of the little ice age in Europe (Fagan 2000).

At the end of eighteenth century, after the plague had receded and as increasing income improved nutrition, housing, and sanitation, life expectancy in Europe and the United States was approximately 40 years (Vallin and Meslé 2009). As Table 5.1 shows, this was a transitional stage at which there were just about as many deaths to children under age 5 as there were deaths at age 65 and over. Infectious diseases (including influenza, acute respiratory infections, enteric fever, malaria, cholera, and smallpox) were still the dominant reasons for death, but their ability to kill was diminishing. Analyses of recently created sets of mortality data have shown, however, that the highest life expectancy recorded anywhere in the world began to go up almost without interruption beginning in about 1800 (Oeppen and Vaupel 2002; Vallin and Meslé 2009). This was led almost exclusively by Scandinavian countries until the 1980s when Japan took over the lead.

Although death rates began to decline in the nineteenth century, improvements were at first fairly slow to develop for various reasons. Famines were frequent in Europe as late as the middle of the nineteenth century—the Irish potato famine of the late 1840s and Swedish harvest failures of the early 1860s are prominent examples. These crop failures were widespread, and it was common for local regions to suffer greatly from the effects of a bad harvest because poor transportation made relief very difficult. Epidemics and pandemics of infectious diseases, including the 1918 influenza pandemic, helped to keep death rates high even into the twentieth century. In August 1918, as World War I was ending, a particularly virulent form of the flu apparently mutated almost spontaneously in West Africa (Sierra Leone), although it was later called "Spanish Influenza." For the next year, it spread quickly around the world, killing more than 20 million people in its path, including more than 500,000 in the United States and Canada (Crosby 1989).

Until recently, then, increases in longevity were primarily due to environmental changes that improved health levels, especially better nutrition and increasing standards of living, not to better medical care:

> Soap production seems to have increased considerably in England, and the availability of cheap cotton goods brought more frequent change of clothing within the economic feasibility of ordinary people. Better communication within and between European countries promoted dissemination of knowledge, including knowledge of disease and the ways to avoid it, and may help to explain the decline of mortality in areas which had neither an industrial nor an agricultural revolution at the time. (Boserup 1981:124–125)

McKeown and Record (1962), who did the pioneering research in this area, and more recently Fogel (2004), argue that the factors most responsible for nineteenth-century mortality declines were improved diet and hygienic changes, with medical improvements largely restricted to smallpox vaccinations. Preston and Haines (1991), though noting the importance of nutrition, have also highlighted the role that knowledge about public health plays in controlling infectious disease:

> In 1900, the United States was, as it is now, the richest country in the world (Cole and Deane 1965:Table IV). Its population was also highly literate and exceptionally well-fed. On the scale of per capita income, literacy, and food consumption, it would rank in the top quarter of countries were it somehow transplanted to the present. Yet 18 percent of its children were dying before age 5, a figure that would rank in the bottom quarter of the contemporary countries. Why couldn't the United States translate its economic and social advantages into better levels of child survival? Our explanation is that infectious disease processes . . . were still poorly understood . . . .(Preston and Haines 1991:208)

Clean water, toilets, bathing facilities, systems of sewerage, and buildings secure from rodents and other disease-carrying animals are all public ingredients for better health. We now accept the importance of washing our hands as common sense, but the important work of Semmelweis in Vienna, Lister in Glasgow, and Pasteur in Paris in validating the germ theory all took place in the mid-nineteenth century—just a heartbeat away from us in the overall timeline of human history. Public health is largely a matter of preventing the spread of disease, and these kinds of measures have been critical in the worldwide decline in mortality. The medical model of curing disease gets much more attention in the modern world, but its usefulness is predicated on the underlying foundation of good public health (Meade and Emch 2010). Cutler and Miller (2005) point to the particularly important role played by the introduction of clean water technology (chlorination and filtration) in cities of the United States in the late nineteenth and early twentieth centuries, the time period when life expectancy made its single biggest jump in U.S. history. This was, of course, a direct application of the germ theory.

Public health improvements, as implied by their name, are viewed as public goods that are paid for societally, rather than individually. Medical care, on the other hand, is still viewed in many parts of the world as an individual responsibility, not a public one. It was not until the early twentieth century in the United States that the health of children came to be seen as a responsibility of the community, rather than just a private family matter (Preston and Haines 1991). Working especially with the school system, this created an atmosphere in which, for example, vaccinations for childhood diseases became widespread. Later on, especially in Europe and Canada, the idea emerged strongly that all aspects of health care, including medical care, should be treated as a public good rather than as an individual affair.

Life expectancy has increased enormously since the mid-nineteenth century. In 1851 in England, the life expectancy for males was only 40 years, and it was 44 for women. At the beginning of the twentieth century, it had increased to 45 for men and 49 for women. But, early in the twenty-first century, life expectancy in the United Kingdom is 80 for men and 84 for women. As you can see in Figure 5.2, this pattern has been closely followed in the United States. In 1850, the numbers in the United States were 38.3 years for males and 40.5 years for females. Referring to Table 5.1, this meant that about 73 babies out of 100 would survive to age 5 and about 29 percent of people born would still be alive at age 65. Figure 5.2 also shows that life expectancy began to increase more rapidly as we moved into the twentieth century and public health measures, in particular, started to have a positive impact. Data for Canada are available only since 1920, but you can see that Canada has always had a slightly higher life expectancy than has the United States.

Looking at Latin America, we can see that prior to the Spanish invasion in the sixteenth century, the area was dotted with primitive civilizations in which medicine

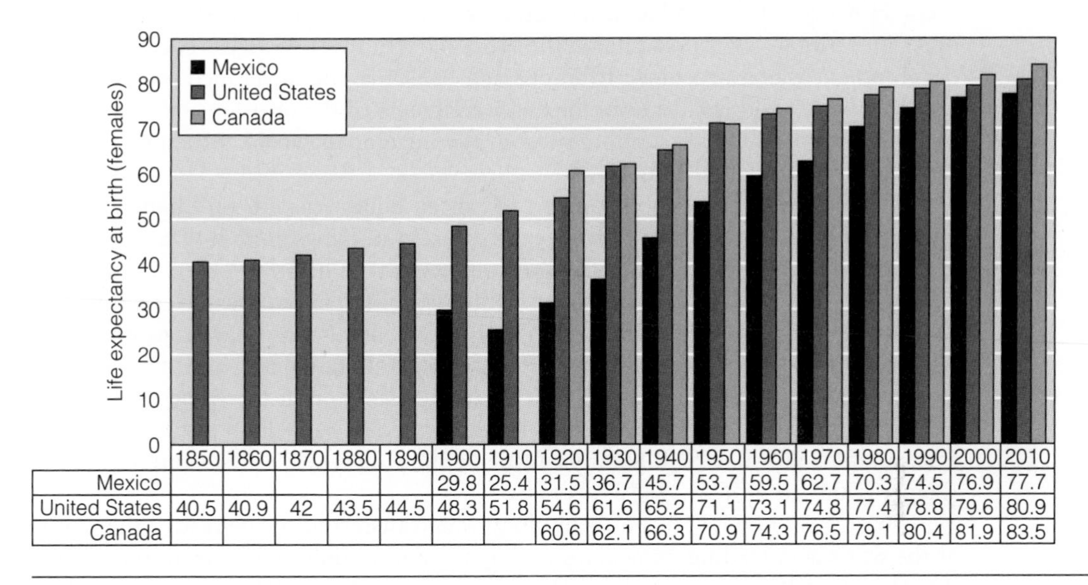

	1850	1860	1870	1880	1890	1900	1910	1920	1930	1940	1950	1960	1970	1980	1990	2000	2010
Mexico						29.8	25.4	31.5	36.7	45.7	53.7	59.5	62.7	70.3	74.5	76.9	77.7
United States	40.5	40.9	42	43.5	44.5	48.3	51.8	54.6	61.6	65.2	71.1	73.1	74.8	77.4	78.8	79.6	80.9
Canada								60.6	62.1	66.3	70.9	74.3	76.5	79.1	80.4	81.9	83.5

Figure 5.2 Life Expectancy Has Improved Substantially in the United States, Canada, and Mexico (Data for Females)

Sources: Data for the United States 1850 through 1970 are from the U.S. Census Bureau, 1975, *Historical Statistics of the United States, Colonial Times to the 1970 Bicentennial Edition, Part I* (Washington, DC: Government Printing Office); Tables B107-115 and B126-135 (data for 1850 through 1880 refer only to Massachusetts); data for Mexico 1900 to 1950 are from Martha Mier y Terán, 1991, "El Gran Cambio Demográfico," *Demos* 5:4-5; Data for Canada 1920 to 1970 are from Statistics Canada, Catalogue no. 82-221-XDE; all other data are from the World Health Organization, Global Health Observatory Data Repository, http://apps.who.int/gho/data/node.main.692?lang=en, accessed 2014.

was practiced as a magic, religious, and healing art. In an interesting reconstruction of history, Bernard Ortiz de Montellano (1975) conducted chemical tests on herbs used and claimed to have particular healing powers by the Aztecs in Mexico. He found that a majority of the remedies he was able to replicate were, in fact, effective. Most of the remedies were for problems very similar to those for which Americans spend billions of dollars a year on over-the-counter drugs: coughs, sores, nausea, and diarrhea. Unfortunately, these remedies were not sufficient to combat most diseases and mortality remained very high (life expectancy less than 30 years) in Mexico until the 1920s, when things started to improve at an accelerating rate. Since the 1920s, death rates have been declining so rapidly that Mexico has reduced mortality to the level that the United States achieved in the 1980s. Thus, in 1920, life expectancy for females in Mexico was 23 years less than in the United States, whereas by 2010 that difference had been cut to 3 years.

World War II as a Modern Turning Point

As mortality has declined throughout the world, the control of communicable diseases has been the major reason, although improved control of degenerative disease has played an increasingly important part. This is true for the less-developed nations of the world today, just as it was for Europe and North America before them. However, there is a big difference between the more developed and less developed countries in what precipitated the drop in death rates. Whereas socioeconomic development was a precursor to improving health in the developed societies, the less developed nations have been the lucky recipients of the transfer of public health knowledge and medical technology from the developed world. Much of this has taken place since World War II.

World War II conjures up images of German bombing raids on London, desert battles in Egypt, D-Day, and the nuclear explosion in Hiroshima. It was a devastating war costing more lives than any previous combat in history. Yet it was also the staging ground for the most amazing resurgence in human numbers ever witnessed. To keep their own soldiers alive, each side in that war spent huge sums of money figuring out how to prevent the spread of disease among troops, including ways to clean up water supplies and deal with human waste, and at the same to work on new ways to cure disease and heal sick and wounded soldiers. Very importantly, for example, World War II brought us penicillin, the world's first "miracle drug" (Hager 2006).

All of this knowledge and technology was transferred to the rest of the world at the war's end, leading immediately to significant declines in the death rate. Thus, it took only half a century in Latin America for mortality to fall to a point that had taken at least five centuries in European countries, as I noted above with respect to Mexico. Countries no longer have to develop economically to improve their health levels if public health facilities can be emulated and medical care imported from richer countries. As Arriaga (1970) noted during the time that this phenomenon was first becoming obvious: "Because public health programs in backward countries depend largely on other countries, we can expect that the later in historical time a

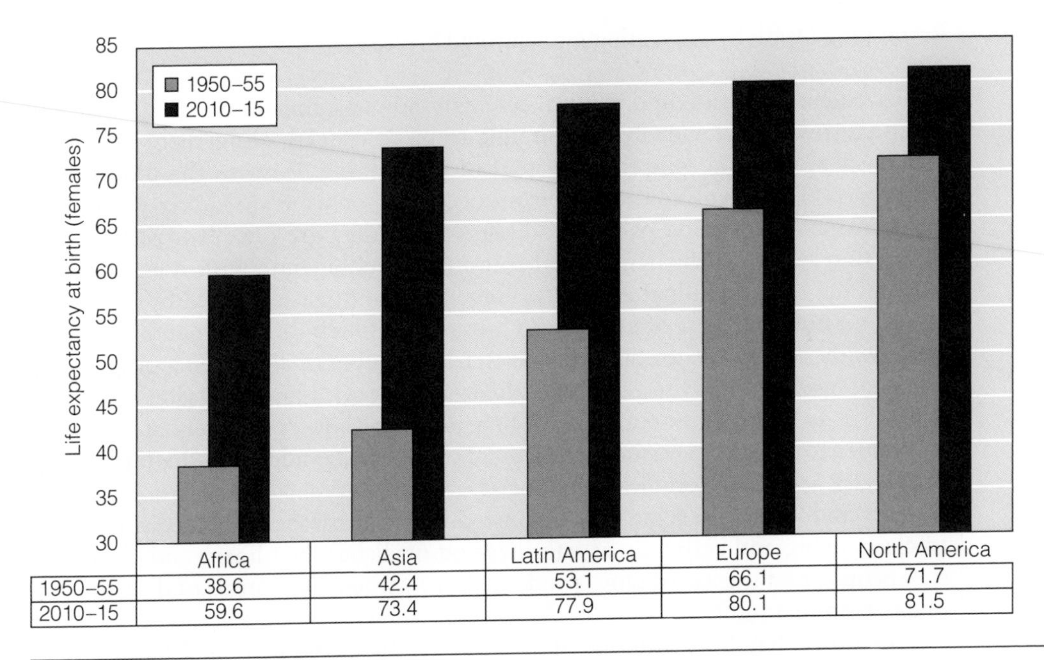

	Africa	Asia	Latin America	Europe	North America
1950–55	38.6	42.4	53.1	66.1	71.7
2010–15	59.6	73.4	77.9	80.1	81.5

Figure 5.3 Regional Changes in Life Expectancy since the End of World War II

Source: Adapted from data in United Nations Population Division 2013, World Population Prospects, the 2012 Revision http://esa.un.org/unpp/ (accessed 2014).

massive public health program is applied in an underdeveloped country previously lacking public health programs, the higher the rate of mortality decline will be." As you can see in Figure 5.3 this applies especially to Latin America and Asia, where improvements in life expectancy have been nothing short of remarkable since the end of World War II.

Progress is not, however, automatic. Sub-Saharan Africa was generally experiencing a rise in life expectancy until being devastated by HIV/AIDS over the past decades, as I will discuss later in the chapter. Eastern Europe in the post-Soviet era experienced a mortality backslide (Carlson and Watson 1990), which pushed life expectancy for Europe as a region below that of North America, as can be seen in Figure 5.3. Data for Russia and Ukraine reveal that life expectancy for males in 2010 was lower than it had been in 1960. Life expectancy had been falling behind other Western nations since at least the 1970s (Vishnevsky and Shkolnikov 1999), and it has been suggested that Russia's health system was unable to move beyond the control of communicable disease to the control of the degenerative diseases, especially related to alcoholism among males, that kill people in the later stages of the epidemiological transition. With the collapse of the Soviet Union in 1989, the health of Russians undoubtedly was further threatened by "political instability and human turmoil" (Chen et al. 1996:526). Life expectancy seems now to have stopped its decline in Russia, but it is not yet clear how quickly (or even, if) it will begin to climb to the levels of western Europe.

Postponing Death by Preventing and Curing Disease

Improvements in health and medical care can only postpone death to increasingly older ages; we are obviously not yet able to prevent death altogether. There are two basic ways to accomplish the goal of postponing death to the oldest possible ages: (1) preventing diseases from occurring or from spreading when they do occur; and (2) curing people of disease when they are sick. Not getting sick in the first place is clearly the ideal route to travel, a route with both communal (public) and individual elements. Prevention of disease is aided by improved nutrition, both in terms of calories and in terms of vitamin and mineral content; clean water to prevent the spread of water-borne disease and to encourage good personal hygiene; piped sewers to eliminate contact with human waste; treatment of sewage so that it does not come back around to "bite" you; adequate clothing and shoes to prevent parasites from invading the body; adequate shelter to keep people dry and warm; eradication of or at least protection against disease-carrying rodents and insects; vaccinations against childhood diseases; use of disinfectants to clean living and eating areas; sterilization of dishes, bed linen, and clothes of sick people; and the use of gloves and masks to prevent the spread of disease from one person to another.

Smallpox has been eliminated as a disease from the world (although there are reportedly still vials in laboratories) as a result of massive vaccination campaigns, and polio is close to being eradicated after a nearly three-decade campaign of worldwide vaccination by the World Health Organization (WHO). Pakistan, Afghanistan, and Nigeria are the only places in the world that still record significant numbers of polio victims each year. In 2013, more than a dozen children were crippled by polio in the civil war-torn country of Syria, and WHO concluded that the infection had originated with someone from Pakistan who had brought it with them to Syria. It is probable that if all 7 billion of us wore sterile face masks for just a few days in succession, we could eliminate several important diseases on a worldwide basis.

Cures for disease range from relatively simple but incredibly effective treatments such as oral rehydration therapy for infants (widely available only since the 1970s), and antibiotics used in the treatment of bacterial infections (widely available only since the 1940s), to the more complex and technology-oriented treatments for cancer, heart disease, and other degenerative diseases, even including organ transplants and stem cell therapies. These high-tech measures include combinations of drug therapy, radiation therapy, and surgery.

This wide range of options available for pushing back death reveals the complexity of mortality decline in any particular population. As Schofield and Reher (1991) noted in a review of the European mortality decline, "There is no simple or unilateral road to low mortality, but rather a combination of many different elements ranging from improved nutrition to improved education" (p. 17). Nonetheless, Caldwell (1986) pointed out that although a high level of national income is nearly always associated with higher life expectancies, the bigger question is whether it is possible for poorer countries to lower their mortality levels. Global experience shows the answer to be yes, and there are several ways to do it.